SHALLOW GROUNDWATER

INTERNATIONAL CONTRIBUTIONS TO HYDROGEOLOGY

18

INTERNATIONAL ASSOCIATION OF HYDROGEOLOGISTS

Shallow Groundwater Systems

Edited by

Peter Dillon
*Center for Groundwater Studies, and CSIRO Land and Water, Adelaide,
Australia*

Ian Simmers
Faculty of Earth Sciences, Free University, Amsterdam, Netherlands

A.A. BALKEMA/ROTTERDAM/BROOKFIELD/1998

Cover photograph: Valley of salt. Photograph by Bill van Aken, CSIRO, Australia.

Published by
A.A. Balkema, P.O. Box 1675, 3000 BR Rotterdam, Netherlands
Fax: +31.10.4135947; E-mail: balkema@balkema.nl; Internet site: http://www.balkema.nl

A.A. Balkema Publishers, Old Post Road, Brookfield, VT 05036-9704, USA
Fax: 802.276.3837; E-mail: info@ashgate.com

ISBN 90 5410 442 2 hardbound edition
ISBN 90 5410 443 0 paper edition (IAH member)

© 1998 A.A. Balkema, Rotterdam
Printed in the Netherlands

Contents

Preface

This volume arose from the recognition that unconfined aquifers are in hydraulic continuity with the unsaturated zone and recharge rates and groundwater quality are intimately related to land use and land management.

Shallow groundwater systems may also be closely related to streams, lakes and wetlands and have a profound impact on both their existence and the associated ecosystems.

Improving our understanding of these interdependencies of shallow groundwater systems with overlying land and surface water relies on integrating knowledge from a range of scientific disciplines, of which hydrogeology is central.

A series of papers was selected from the XXV Congress of IAH 'Water Down Under '94' Adelaide, Nov 1994, with the theme 'Management to Sustain Shallow Groundwater Systems'. Many of these selected papers have been substantially revised and updated. These papers highlight our understanding of unsaturated zone processes and surface water-groundwater interaction. We also note that the ability to model such systems is important in developing confidence in managing such systems where decisions may have long-term and far-reaching consequences. This volume is therefore dedicated to improving our understanding of, and our ability to model, shallow groundwater systems.

Acknowledgements:
The editors are grateful for help received from reviewers of the papers presented: T.M. Addiscott, D. Armstrong, W. Bond, T. Chapman, P. Cook, G. Davis, P.J. Dillon, R. Evans, F Ghassemi, L.F. Konikow, J.W. Mercer, K. Narayan, A.G. Reynders, D. Rolston, P. Ross, R. Salama, V.O. Snow and I. White.

Our thanks also go to Heather Bajcarz, of the Centre for Groundwater Studies, who assisted with correspondence and reformatting of a number of papers.

Peter Dillon
Centre for Groundwater Studies, and CSIRO Land and Water, Adelaide, Australia

Ian Simmers
Faculty of Earth Sciences, Vrije Universiteit, Amsterdam, Netherlands

Invited review

CHAPTER 1

Shallow groundwater systems

LLOYD R. TOWNLEY
Centre for Groundwater Studies, CSIRO Land and Water, Private Bag, PO Wembley, Australia

ABSTRACT: Shallow groundwater systems are those that occur nearest the earth's surface, bounded by a water table and an unsaturated zone above, and by bedrock or deeper aquifers below. Shallow groundwater systems are more likely than deeper systems to be in contact with surface water bodies, thus the interaction between surface water and groundwater, from both physical and biogeochemical points of view, can significantly influence shallow aquifers. This paper provides an outline of a selection of papers presented at the *Water Down Under* conference held in Adelaide, Australia, in November 1994. The papers are presented in the context of surface water – groundwater interaction, this being either direct contact between aquifers and surface water bodies or indirect interaction through the unsaturated zone. Many of the papers focus on modelling of flow and contaminant transport in aquifers and the unsaturated zone, with consideration of the effects of heterogeneity, and in one case, the effects of density.

1 INTRODUCTION

Shallow groundwater systems are those which occur nearest to the earth's surface. They are bounded above by a water table, or more generally by an unsaturated zone between the water table and the land surface. They are bounded below by impermeable barriers, by aquitards, or by deeper groundwater systems which are recharged upgradient and discharge downgradient of the more localised shallower groundwater systems. This definition is hardly precise, but provides a workable definition for the purposes of this paper.

Because of their proximity to the earth's surface, shallow groundwater systems are more likely than deeper systems to be in contact with surface water. An understanding of shallow groundwater systems therefore requires an understanding of the interaction between surface water and groundwater, from both physical and biogeochemical points of view. Because the issue of surface water – groundwater interaction is so important for shallow groundwater systems, a group of papers presented at *Water Down Under* has been chosen for this special volume to address this theme.

As with many specialised areas of hydrology, the study of shallow groundwater systems relies on an appropriate set of tools. In order to analyse and predict the behaviour of shallow groundwater systems, it is both useful and necessary to use models of flow and solute transport, these models being conceptual and mathematical, and in some cases implemented in the form of widely available computer codes. It is for this reason that a second group of papers presented at *Water Down Under* has also been chosen for this special volume.

The purpose of this paper is to introduce the issues of surface water – groundwater interaction and flow and solute transport modelling in the context of shallow groundwater systems. The paper draws on the author's general knowledge of important issues and refers, where appropriate, to the papers chosen for this special volume.

2   SURFACE WATER – GROUNDWATER INTERACTION

The earth's surface is covered by a variety of vegetated and unvegetated areas, and by surface water bodies of various shapes and sizes. All of these surface types are potentially underlain by shallow groundwater systems. The topic of surface water – groundwater interaction therefore includes a wide variety of interactions, from recharge and discharge processes through the unsaturated zone, to direct connections between surface water bodies and aquifers.

To a large extent, surface water – groundwater interaction controls the behaviour of all groundwater systems. In a simple model of the hydrologic cycle, rainfall is balanced by evapotranspiration, surface runoff, and recharge to or discharge from shallow groundwater systems. The location of recharge and discharge zones depends largely on topography, or position in the landscape, thus in general recharge zones tend to be in uplands and discharge zones tend to be in topographic lows, perhaps nearer to surface water bodies. Aside from the abundance of rainfall, the magnitude of recharge is controlled largely by the hydraulic characteristics of the unsaturated zone, as well as by vegetation, depth to water table and the presence of pathways for preferential flow. Once recharge has occurred, subsequent discharge usually occurs at or near surface water bodies.

Most natural surface water bodies exist in topographic lows and act as focal points in the hydrologic system, such that water on the surface and groundwater is attracted towards them. Surface water bodies such as rivers and streams are often used to define the boundaries of local scale studies of groundwater systems. Some artificial surface water bodies, and some natural ones as well, are elevated above regional groundwaters, and these water bodies can act as sources of recharge. In general, the direction of fluxes between a surface hydrological system and shallow groundwater system depends on hydraulic gradients, which in turn depend on topography and the balance of flows in the whole hydrological system. The magnitude of fluxes depends on saturated and unsaturated hydraulic properties of the aquifer and unsaturated zone. Whether there is direct connection with a surface water body, or indirect connection through an unsaturated zone, the hydraulic gradients and hydraulic properties in these zones of surface water – groundwater interaction control the flows between the two systems.

2.1   *Recharge-discharge processes through the unsaturated zone*

Movement of water through surface soils is a complex dynamic process that depends on the physical and chemical properties of the soil, the distribution and types of vegetation and on atmospheric conditions which cause rainfall and evapotranspiration. Infiltration, exfiltration, redistribution and evapotranspiration are all influenced by variability in the properties of the soil, vegetation and climate, at all scales in space and time.

Kookana and Naidu (this volume) provide a useful discussion of the importance of heterogeneity in physical, chemical and biological properties of soil in controlling contaminant transport through unsaturated soil profiles. Such heterogeneities are not independent of each other, and it is becoming increasingly clear that variability in some chemical or (micro)biological characteristic of a soil can compound or oppose the effects of variability in some physical characteristic.

Snow et al. (this volume) demonstrate by experiments and modelling that the initial moisture content of the surface layer of a soil greatly affects subsequent solute movement, particularly in the short term. The authors argue that more understanding is needed of the processes influencing movement of fertilisers and contaminants into and out of immobile water near the soil surface.

Smettem et al. (this volume) present laboratory and field results using a twin disc infiltrometer for measuring hydraulic conductivity. The field experiments were conducted on a well aggregated calcareous loam under permanent pasture in Hampshire, UK. The authors show that the assumption of one-dimensional flow at early times near a single disc infiltrometer may lead to erroneous estimates of hydraulic conductivity. Furthermore, if infiltration is governed by capillary flow effects, even a twin disc infiltrometer should not be used to estimate hydraulic conductivity.

Recharge and discharge processes vary at all scales, and a major issue is extrapolating our knowledge from point scale to regional scale. Modern geographical information systems (GIS) provide tools for extrapolating, as do computer models of groundwater flow. But regional computer models need to be coupled with physically-based models of recharge and discharge processes, and such coupled models are not always readily available.

Water quality in shallow aquifers is influenced by land use activities of all kinds. One methodology for land use planning which has achieved some degree of popularity is that of vulnerability assessments. Barber et al. (this volume) compare the DRASTIC procedure, developed in the USA, with a simpler region-specific spatial modelling approach, applied to a region in New South Wales, Australia. They show that both methods underestimate vulnerability to both distributed and point source pollution by nitrate, and conclude that there is a need to define uncertainties in vulnerability assessments, to avoid the potentially overly prescriptive use of vulnerability maps in the future.

In some parts of the world, basic hydrogeological characteristics of significant areas are not yet known. Taylor and Howard (this volume) study recharge mechanisms to shallow aquifers in two catchments in Uganda. Water supplies have traditionally been drawn from fractured bedrock aquifers, but the overlying regolith (weathered zone) also acts as a significant source of groundwater. The hydraulic connection between the regolith and the bedrock aquifers has been studied by short- and long-term pumping tests, and also using inorganic and isotope hydrochemistry. Average recharge in one catchment, with lower relief and higher rainfall, is estimated to be ten times higher than in the other; an extensive marsh in the first catchment (known as a dambo, and occupying 7% of the catchment area) is shown not to contribute significantly to groundwater recharge.

Changes in land use are known to have significant influences on recharge in semi-arid parts of Australia. Prathapar et al. (this volume) predict the impact of clearing of natural vegetation in a region of more than 20,000 km^2 in south-western New South Wales, Australia. They use groundwater flow models coupled to an economic model to predict the long-term impacts of clearing on salinisation of the land and increase in the salinity of the Murray River which receives groundwater discharge. Recharge to the water table is assumed to depend on infiltration at the surface, initial depth to water table and time since clearing.

2.2   *Direct connection between surface water bodies and groundwater*

Surface water bodies occur is many forms, for example as wetlands, lakes, ponds, rivers, streams, canals and drains. They occur in the lower parts of the landscape, either in relatively closed depressions capable of storing significant quanitities of water or in gently sloping features capable of conveying water across the landscape. Surface water bodies can be classified as discharge water bodies, if they receive groundwater over the whole of their bottom surface, as recharge water bodies, if they provide recharge to an underlying aquifer over the whole of their bottom surface, and as flow through water bodies, if they receive groundwater over part of their bottom surface and recharge an aquifer over the remaining part (Born et al., 1979). A more detailed classification defined on the basis of groundwater flow patterns is provided by Nield et al. (1994), and a useful climatic and physiographic classification is provided by Winter (1992).

De Vries (this volume) relates the density of stream networks and the distribution of channel size in the Netherlands to the nature of seasonal rainfalls. The results are based on the premise that a stream network in a shallow sandy aquifer provides a mechanism for removal of groundwater draining to the streams. From a geomorphological point of view, it is therefore natural to expect that the depth and spacing of streams are determined by some equilibrium with local geological and climatic conditions.

Batelaan et al. (this volume) provide an interesting case study in Belgium in which groundwater discharge to the upper part of a stream network is shown to originate as recharge in another surface water catchment, on the other side of a topographic divide. The results are supported by remote sensing and vegetation mapping, geochemical analyses and steady-state quasi-three-dimensional groundwater flow modelling with capture zone analysis.

3   FLOW AND SOLUTE TRANSPORT MODELS

Models provide useful tools for the study of shallow groundwater systems. Models can be of many types. Nearly all models depend on an initial conceptual model: A conceptualisation of the geometry of interest and the important processes taking place. The next step is frequently a representation of the conceptual model in mathematical terms, perhaps in the form of differential equations (representing water or solute balances), or perhaps in the form of a lumped model or a statistical model. Mathematical models can often be solved by many different methods, thus the same mathematical model might be considered to be different if it were solved analytically, or by finite difference or finite element or boundary integral element methods. A computer code, which is a particular implementation of a particular solution technique for a particular mathematical formulation, is also often described as a model. Finally, a particular computer code, when applied to a particular field site and calibrated using field data is also described as a model of that site. All these kinds of models are potentially useful, as will be seen in the examples below.

3.1 *Models of groundwater flow*

The topic of groundwater flow modelling in shallow aquifers is now a mature field in hydrology, and there are many available codes which can be used relatively easily to predict flow in two dimensions in plan, in multi-layered quasi three-dimensional systems and even in three dimensions. On the other hand, there remain many weaknesses in available codes. Few codes contain algorithms for estimating recharge and evapotranspiration on a regional scale, and users are required to estimate these quantities independently, external to the groundwater flow model, and to supply them as inputs to the model. Few codes include robust links to surface water bodies, whether they are narrow linear features like rivers or streams, or larger closed bodies like lakes and wetlands. A class of problems which is not commonly addressed in available codes is that of moving boundary problems, such as when the shoreline of a lake advances and retreats seasonally; there are no readily available user-friendly codes which can handle this type of problem in plan or in three dimensions.

Haagsma and Johanns (this volume) propose a method for loosely coupling MODFLOW (McDonald & Harbaugh, 1984), as a typical groundwater flow model, with DUFLOW, a one-dimensional model of unsteady flow in an open channel. They argue that tightly coupled models, where both groundwater and surface water flows are computed in the same code, are undesirable. They then define a communications interface which allows the two codes to run in parallel, continuously passing results to each other as needed. Models of this kind may be useful in the context of decision support systems for water management.

3.2 *Models of contaminant transport in aquifers*

Contaminant transport has been the focus of most research in groundwater hydrology since the 1980s. It is therefore not surprising that there is much interest in the development of models of contaminant transport. Research has focused on the improvement of numerical techniques for accurately solving the advection-dispersion equation, and on methods for incorporating the effects of heterogeneity in aquifer hydraulic conductivity into such transport models. There is and has long been considerable interest in karst aquifers, especially in Europe where many significant water supplies occur in karst. Many research projects have included tracer tests, where dyes or other passive substances are used to trace the movement of groundwater in aquifers.

Ptak (this volume) accounts for the effect of heterogeneity in hydraulic conductivities on the analysis of tracer experiments using a three-dimensional Monte Carlo stochastic transport model. An indicator approach is used to evaluate the geostatistical structure of the aquifer at the Horkheimer Insel in South Germany, and the conditional sequential indicator simulation method is used to generate three-dimensional realisations of hydraulic conductivity. Groundwater flow and advective transport are then simulated using MODFLOW (McDonald & Harbaugh, 1984) and MODPATH (Pollock, 1989), thus allowing interpretation of a multilevel forced gradient tracer test with fluoresceine.

Wang and Apperley (this volume) illustrate a technique for predicting contaminant transport in one and two dimensions in stratified porous media, based on a Laplace transform technique. This technique removes the difficulties associated with time-stepping in traditional models, and works well as long as numerical inversion of the Laplace-transformed variables is robust.

Maloszewski et al. (this volume) apply two models to the interpretation of artificial tracer experiments in a karst aquifer system north of Graz in Austria, primarily using the fluorescent dye uranine, but also using bromide and chloride. The first model is based on several parallel pathways, each based on a classical advection-dispersion equation, with a different volume, velocity and dispersivity. The second is a double porosity model, with flow in several parallel fractures, and diffusive transport into a porous matrix with sorption. Both models are calibrated to match the field data, but there are insufficient data to choose either model in preference to the other. An attempt is made to calibrate a classical model based on a single pathway, using the method of moments, but this model cannot be fitted to the data.

Laslett and Davis (this volume) describe the use of HST3D (Kipp, 1987) to the interpretation of a tracer test in a shallow aquifer near Perth, Western Australia, where large changes in the water table elevation occurred. The authors have modified HST3D by including more efficient equation solvers, and have also developed a post-processor to facilitate its use. Rather than representing the aquifer as an unconfined aquifer, they choose to represent it as a confined aquifer and to distribute the storage due to specific yield over several layers near the top of the aquifer. In general, their attempts to simulate the tracer test were unsuccessful, perhaps due to heterogeneity which was not represented in their model.

3.3 *Models of contaminant transport in the unsaturated zone*

Transport of contaminants in the unsaturated zone is important because the movement of contaminants through this zone to the water table often provides the source of contamination for regional aquifers. Just as predicting flow in the unsaturated zone is inherently more difficult than predicting flow in aquifers, contaminant transport in the unsaturated zone is similarly complex.

Binning and Celia (this volume) present a new two-dimensional model of multiphase flow and contaminant transport in the unsaturated zone. The phases considered are water and air, and the finite element model includes retardation, reactions and both equilibrium and non-equilibrium mass transfer between phases. The authors argue that it is important to include heterogeneity when modelling water and air movement in the unsaturated zone, and that because of the need to correctly represent the balance between capillary and gravity forces, it is difficult to scale up the problem in either space or time.

Bond et al. (this volume) validate a one-dimensional model of flow and transport in the unsaturated zone, by showing that the model can succesfully predict bromide concentrations during a six-month field trial without any attempt to calibrate the model against the data. The model uses data on soil properties, rainfall and evaporation which were obtained by independent measurements at a field site near Wagga Wagga in New South Wales, Australia. The model is based on Richards' equation, and was run with molecular diffusion only and dispersion set to zero. This study provides some confidence that classical models can indeed be applied to predict movement of water and contaminants in at least some field situations.

3.4 *Models of density-coupled flow and transport*

In some situations it is necessary to take into account the influence of transported contaminants and/or heat on the density of groundwater. When the effects are large enough to influence the directions of flow of groundwater, the flow is known as convection. Examples occur in coastal situations near a seawater wedge, above salt domes, beneath salt lakes and in regions which are geothermally active.

Narayan and Simmons (this volume) study the development of convection cells beneath saline disposal basins in the Murray-Darling Basin in Australia. They apply SUTRA (Voss, 1984) to a two-dimensional cross-section through a saline disposal basin, and illustrate the differences between convective flow patterns and rates of transport of salt as a function of Rayleigh and Nusselt numbers.

4 DISCUSSION

Most of the processes involved in the movement of water and contaminants in shallow aquifers are probably well understood, for most practical purposes. However, there will always be issues that require further investigations, especially when coupled physical, chemical and biological processes are involved, and also where the local characteristics of soils, vegetation and climate make the hydrology of a region apparently different from other areas which are well understood. Coupled reactive processes in heterogeneous aquifers, in both the unsaturated and saturated zones, are very active areas of research (Dracos & Stauffer, 1994). There is still a lack of reliable and affordable measurement techniques, e.g. for direct determination of recharge and contaminant fluxes, particularly where preferential flow paths may be significant. Preferential flow paths may result in infiltration which cannot be adequately described by Richards' equation, in spite of the fundamental validity of that equation in homogeneous media.

The role of surface water bodies in influencing shallow aquifers, and vice versa, is a topic ripe for research. Considering physical processes alone, most of our knowledge is based on steady or slowly varying systems, even though it has been observed that many systems experience significant dynamics at the time scale of individual storm events or in the context of seasonal variations. As water levels change in shallow water bodies, the areal extent of the water surface can change significantly, and this can affect not only the rate but also the direction of regional groundwater flows. Groundwater capture zones may affect the water quality of receiving surface water bodies but, conversely, water quality in surface water bodies may also affect groundwater quality in the release zones from the water bodies. Water quality in meandering streams can be significantly affected by biogeochemical processes in alluvial aquifers along the course of the streams, as surface waters recharge the aquifers and the aquifers discharge to the streams further downstream.

A dilemma in hydrology is that although the processes involved are conceptually simple, there are so many variables involved that a rigorous dimensional analysis of real problems in real heterogeneous aquifers needs a very large number of non-dimensional ratios to define the system behaviour. While problems in fluid mechanics can be presented in terms of well-known quantities like Reynolds, Froude, Schmidt, Strouhal and Rayleigh numbers, for example, the number of widely accepted controlling non-dimensional ratios in hydrology is small. There is very little research in stochastic groundwater hydrology which uses the concepts of dimensional analysis to allow extrapolation of results from one heterogeneous field situation to another.

Most hydrologists agree on the value of real data from the field to validate models and to support their use in future predictions. However the majority of case studies are presented in physical units, with no attempt to non-dimensionalise the results so that they can be applied to a field site of another scale. There remains an opportunity for researchers to develop scaling laws which are relevant in groundwater hydrology, such that the effect of geometric (and all other) scaling on all relevant processes can be predicted (e.g. Kalma & Sivapalan, 1995).

Models of flow and contaminant transport in aquifers and the unsaturated zone are tools in the hands of experienced hydrologists. There are many models or codes applicable to shallow groundwater systems, but at the same time there are not nearly enough. Hydrology is different from most other fields in science and engineering, in that software tools remain relatively immature. By far the majority of codes written by groundwater hydrologists to solve specific problems are used only by the authors of the software, and mostly in a university environment. The majority of codes that are widely used are those that have been developed and popularised by staff of the US Geological Survey. Distribution of such software at a nominal charge has produced many benefits to users, but this practice has generally limited the capacity of others in the private domain to develop and support software of a commercial quality comparable to that available in other fields of science and engineering.

Many recent advances in our understanding of the role of heterogeneity (stochastic groundwater hydrology), in our knowledge of inverse problems and model calibration, in contaminant transport with multiple phases and coupled chemical and biological processes, have not yet been implemented in widely available, let alone robust and user-friendly computer codes. There are clear opportunities for further development of codes as tools for groundwater hydrologists, specifically in the study of shallow groundwater systems, but there remains a question of how, and by whom, such codes will be written. The preceding remarks support the view that there are many reasons why further research on shallow aquifer systems, and further development of tools for studying such systems, are necessary.

Many countries today are operating in socio-economic climates in which there is increasing pressure on science and a tendency for reductions in research funding. At these times, it is more important than ever that government agencies, managers and researchers understand the nature of their interdependence. Today's managers of land and water resources experience uncertainty on a day-to-day basis as they are faced with new and difficult questions. Not all problems faced by managers need new research. However, managers could benefit from closer links with researchers, if only to assist in the interpretation of published literature which can be found in vast quantities using abstracts published on CD ROMs.

In both the short and long-term, in all parts of the world, the hydrological community will need an ongoing supply of experts. Research is the primary means for students to develop real expertise and a grasp of existing published literature. The hydrological community must argue for sustained research in order simply to maintain access to our existing knowledge base.

REFERENCES

Barber, C., Bates, L.E., Barron, R. & Allison, H. 1998. This volume 87-96.

Batelaan, O., de Smedt, F., de Bekker, P. & Huybrechts, W. 1998. This volume 75-86.

Binning, P. & Celia, M.A. 1998. This volume 157-167.

Bond, W.J., Smith, C.J. & Ross, P.J. 1998. This volume 63-73.

Born, S.M., Smith, S.A. & Stephenson, D.A. 1979. Hydrogeology of glacial-terrain lakes, with management and planning applications. *J.Hydrol.* 43: 7-43.

De Vries, J.J. 1998. This volume 53-62.

Dracos, Th. & Stauffer, F. 1994, (eds). Transport and Reactive Processes in Aquifers. *Proceedings of the IAHR/AIRH Symposium on Transport and Reactive Processes in Aquifers, Zurich, Switzerland*, 11-15 April, 1994, 590pp. Rotterdam: Balkema.

Haagsma, I.G. & Johanns, R.D. 1998. This volume 143-156.

Kalma, J.D. & Sivapalan, M. 1995, (eds). *Scale Issues in Hydrological Modelling*, 489pp. New York: John Wiley & Sons.

Kipp, K.L. 1987. HST3D: a computer code for simulation of heat and solute transport in three-dimensional ground-water flow systems. USGS Water-Resources Investigations Report 86-4095.

Kookana, R.S. & Naidu, R. 1998. This volume 15-27.

Laslett, D. & Davis, G.B. 1998. This volume 207-219.

Maloszewski, P., Benischke, R., Harum, T. & Zojer, H. 1998. This volume 177-190.

McDonald, M.G. & Harbaugh, A.W. 1984. A modular three-dimensional finite-difference ground-water flow model. USGS Open-File Report 83-875.

Narayan, K.A. & Simmons, C.T. 1998. This volume 221-232.

Nield, S.P., Townley, L.R. & Barr, A.D. 1994. A framework for quantitative analysis of surface water – groundwater interaction: Flow geometry in a vertical section. *Water Resour. Res.* 30(8): 2461-2475.

Pollock, D.W. 1989. Documentation of computer programs to compute and display pathlines using results from the US Geological Survey modular three-dimensional finite-difference groundwater flow model. USGS Open-File Report 89-831.

Prathapar, S.A., Williams, R.M. & Punthakey, J.F. 1998. This volume 115-126.

Ptak, T. 1998. This volume 129-141.

Smettem, K.R.J., Ros, P.J., Haverkamp, R., Parlange, J.Y. & Gregory, P.J. 1998. This volume 41-51.

Snow, V.O., Clothier, B.E. & Tillman, R.W. 1998. This volume 29-39.

Taylor, R.G. & Howard, K.W.F. 1998. This volume 97-113.

Voss, C.I. 1984. SUTRA: A finite-element simulation model for saturated-unsaturated fluid density-dependent groundwater flow with energy transport or chemically reactive single-species solute transport. USGS Water-Resources Investigations Report 84-4369.

Wang, J.C. & Apperley, L.W. 1998. This volume 169-176.

Winter, T.C. 1992. A physiographic and climatic framework for hydrologic studies of wetlands, In: Aquatic ecosystems in semi-arid regions: Implications for resource management. NHRI Symposium Series 7, Environment Canada, 127-148.

Unsaturated-saturated zone processes

CHAPTER 2

Vertical heterogeneity in soil properties and contaminant transport through soil profiles

R.S. KOOKANA & R. NAIDU
Cooperative Research Centre for Soil and Land Management, Glen Osmond, Australia

ABSTRACT: Soils are highly heterogeneous in physical, chemical and biological properties; the properties which collectively determine the fate and behaviour of a contaminant in the soil profile. To describe transport behaviour of contaminants in soils a variety of solute transport models have been proposed in recent years. These models vary in conceptual approach and degree of complexity; however, they rarely include the scale of heterogeneity of soil properties encountered in the field. Whilst some processes are usually considered in a greater detail during simulation of solute transport, the soil profile is generally considered as uniform in terms of flow and other soil physical and chemical properties. Given that the behaviour of numerous chemicals is governed by soil properties such as soil organic matter, pH, clay content, permeability, etc. which vary markedly with depth, it is important to establish both the depth-wise distribution of these properties within a soil profile and the functional description of the behavioural dependence of the contaminants on these properties. Most such relationships are either not available or have not yet been built into the models. This paper discusses the chemical and physical heterogeneous nature of a soil profile and highlights the need of their incorporation in models as well as their importance in controlling contaminant behaviour.

1 INTRODUCTION

Leaching of contaminants through soil profiles can affect the quality of surface and groundwaters, and therefore modelling of the transport behaviour in soils is useful. A large number of transport models, varying in conceptual approach, degree of complexity and in their coverage of processes of the soil-contaminant interactions, have been developed in recent years (Wagenet & Rao, 1989). To model the behaviour of a contaminant in a complex and dynamic system such as soil is inherently a difficult task. This is so because a contaminant following its contact with soil, can undergo several biogeochemical processes such as sorption-desorption, transformation/degradation, volatilization and leaching. The dynamics of the contaminants in soil systems are further complicated by the different time scales of the interacting processes (Jury & Roth, 1990). Ideally, a comprehensive transport model should cover all the soil-chemical interactions and provide accurate predictions of the concentrations resulting from these interactions. In practice, however, this may be an unsurmountable task and therefore all these processes are not or cannot be included in the model in a comprehensive manner and often description of the processes is simplified. Furthermore, the enormous heterogeneity of soil systems poses additional difficulties.

Most of the existing models are based on traditional deterministic approaches, which may be of limited applicability in the field due to the inherent spatial variability of soil properties both in horizontal and vertical directions. One approach used in recent years to deal with soil variabilities is to formulate transport process in terms of an integral property of the soil (Jury & Roth, 1990), a resultant of all heterogeneities affecting the transport process. While the horizontal variability has received increasing attention in recent years, the vertical variation of soil properties has only recently been recognised to have a major impact on contaminant transport (Russo, 1991). Natural field soils are, however, generally more homogeneous in the horizontal direction than vertical, due to the very process of soil genesis (Ellsworth & Jury, 1991). Even models accounting for the lateral variability of flow properties of soils by treating solute transport as a stochastic process (Bresler & Dagan, 1981; Jury et al., 1986), assume the soil to be composed of a series of isolated stream tubes, homogeneous along the direction of flow. In reality, however, not only the physical properties (such as texture, structure, organic carbon and permeability) but also chemical and biological properties (such as pH, exchangeable cations, microbial activity, etc.) change markedly with depth in a soil profile. These vertically changing properties not only affect sorption, degradation and contaminant transport through soil profiles, but also impact on the soil hydraulic properties and longitudinal dispersion of a contaminant.

Undoubtedly, vertical heterogeneity introduces extreme complexity into the field description of solute transport and will require a three-dimensional approach to adequately understand the impact of such heterogeneity on solute transport through soils (Russo, 1991; Ellsworth & Jury, 1991). This paper discusses the importance of some soil profile characteristics in modifying the leaching behaviour of contaminants and thus highlights the need for incorporation of these aspects in future solute transport models. The relationship between vertical variation in soil properties and contaminant behaviour in a soil profile have been illustrated through some examples.

2 SOIL PROFILE AND VERTICAL DISTRIBUTION OF SOIL PROPERTIES IN THE PROFILE

Far from being homogeneous, soil profiles consist of horizontal layers or 'horizons' from surface to bedrock, each differing with respect to organic and inorganic composition of the soil material. Usually, there are three sets of horizons in a soil profile. The A horizon, the uppermost soil layer, is a zone of net depletion and may be rich in organic matter. The middle layer, known as the B horizon, is the region of accumulation where increased clay content and sometimes organic matter may be found. The C horizon is a heterogeneous accumulation of unconsolidated material from which the soil has formed (Fig. 1). Each soil layer can be further subdivided into other zones based on morphological, chemical and physical (textural) characteristics. The physical properties of the soils, textural classes in particular, vary markedly with depth. Generally, the clay content is lowest in the surface horizon rising to its highest value in the subsurface B horizon (Fig. 1). In this horizon extractable Fe, Al and Si levels are also highest. In addition to the elevated levels of metal ions, certain soils formed under acid weathering regimes contain high concentrations of organic carbon relative to the surface. Not only the soil properties but the pedoclimate can also vary significantly in a soil profile. Pedoclimate reflects the moisture regime and redox conditions at microsites within each soil layer in the unsaturated zone of the soil profile.

Depending on the soil type, pH which reflects the chemical reactivity of soils, may vary widely with increasing depth within a profile. With the exception of self-mulching Black Earths (Vertisols), the pH of many of the Australian soils increases with increasing depth (Fig. 1). For example, in sodic soils, that comprise approximately 33% of the Australian land mass (Northcote & Skene, 1972), soil pH can range from < 4.5 in the surface 25 mm soil to > 7.3 in the sub-surface horizons. In the profiles that are dominantly sodium rich, the pH of the sub-surface horizons may exceed 8.5. In the self-mulching soils, however, there is little variation in the pH and other soil properties with depth.

Unlike soil pH, organic carbon is usually highest in the A horizon. Depending on the soil type it may range from < 0.2% to 10% in the surface soils to < 0.1% in the subsurface horizons, except some soils (Histosols) where organic carbon in surface soils would exceed 10%. The higher values are usually associated with soils under long term pasture and native forest. The dissolved organic carbon (DOC) is highest in the soil solution/seepage waters of the uppermost layers of the soils rich in organic material (Fig. 2). It decreases rapidly towards the mineral soil horizons. The decrease in DOC with depth may be due to a decrease in soil organic matter content, adsorption of the organic substances onto clay minerals, formation of organo-mineral compounds or due to precipitation, flocculation and formation of insoluble complexes.

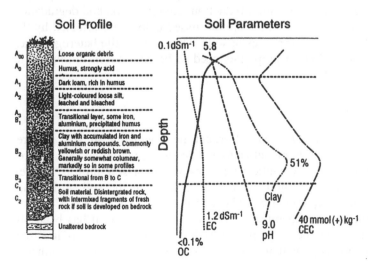

Figure 1. A schematic representation (not to scale) of soil profile and depthwise distribution of key soil properties.

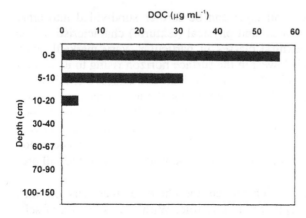

Figure 2. Variation in soil solution concentrations of dissolved organic carbon (DOC) in a sodic xeralf soil profile.

Since the cation exchange capacity (CEC) and total exchangeable cations are functions of clay, pH, organic matter, etc., it is often difficult to generalise the trend within a soil profile. Regardless, in soils which are rich in organic carbon and high in pH, the CEC of surface layers is usually higher than in the subsurface horizons. In those soils which are high in clay content and contain expansive clay minerals in the sub-surface horizons (e.g. xeralfs), CEC increases markedly with increasing depth. Figure 1 shows a typical CEC trend for the duplex soils (e.g. natraqualf) commonly found in agricultural areas. Concomitant with the increase in CEC there is also a tendency for the salt content to increase with depth.

3   IMPLICATIONS OF SOIL PROFILE HETEROGENEITY ON CONTAMINANT BEHAVIOUR

The transport of toxic metals and other organic contaminants in the subsurface horizons is controlled not only by the hydraulic processes but also by the interactions between water, soil and solid rock fragments via adsorption-desorption, dissolution-precipitation, acid-base reactions, oxidation-reduction, complexation and degradation processes. These processes, some of which are discussed below, are affected by the physical, chemical and biological properties of soils. Thus soil profile heterogeneities of physical, chemical and biological properties need to be considered in relation to contaminant transport through soils.

3.1   *Physical heterogeneity*

Clay content and hydraulic conductivity. Vertical heterogeneity in soil texture, structure, bulk density, nature of dominant cations on exchange complex, etc. can result in significant variations in hydraulic properties within a soil profile. Duplex soils, characterised by a sharp textural contrast at a depth in the soil profile, are common in Australia. The sharp increase in clay content in the subsurface B horizon renders it very different physico-chemical properties. For example, it has been observed that the hydraulic conductivity of clay-rich horizons can be so low that perched water table conditions prevail and lateral rather than vertical flow dominates.

During experiments on the leaching of pesticides dibromochloropropane (DBCP) in fallow fields of Georgia, and aldicarb in potato fields of Florida, lateral movement was observed (cited from Helling & Gish, 1986). In both cases the fields were located on sandy soils underlain by an impervious clay layer. Helling and Gish (1986) also discussed several examples where groundwater pollution by pesticides correlated well with the permeability of soil layers or geological formations (such as karst-carbonate aquifers, secondary aquifers protected by clay layers, etc.).

Sodicity, solution composition and hydraulic conductivity. The permeability of a soil is also significantly influenced by the ionic composition of the soil solution and the solid phases, especially in sodic duplex soils where the salt concentration and sodicity generally increase with depth. In such soils, the salt concentration difference between the original saline/sodic soil solution and the percolating fresh water can cause dispersion of clay minerals at the interface between the resident and incoming solutions, thereby clogging both macro/micropores and reducing the permeability.

Heterogeneities of hydraulic properties – field observations. Many workers have observed significant vertical heterogeneity in hydraulic properties during field experiments. For example, Sudicky (1986) measured hydraulic conductivity of each 5 cm layers of a 2 m long core from the Borden aquifer and found the conductivity values to range from 6×10^{-4} to 2×10^{-2} cm s^{-1} – a 30 fold variation. Similarly at a contaminated site in Adelaide, we noted the hydraulic conductivity to range from $< 2 \times 10^{-6}$ to 2×10^{-4} cm s^{-1}, within a vertical distance of 1 m (at 8-9 m depth) because of the differences in soil texture and chemical composition. The hydraulic conductivity of clay layers at this site fell in the range of that for the artificial liners used at sites for waste disposal and therefore the contaminant movement at the site would be restricted to mainly lateral flow or diffusion through the clay layer.

From a limited number of field experiments conducted on solute transport in the vadose zone, it has become clear that physical heterogeneity can cause large spreading of the solute distribution in the profile, so-called macrodispersion. Depth fluctuations in longitudinal dispersivity due to vertical heterogeneity and compression-expansion of tracer plumes have been observed (Butters et al., 1989; Ellsworth et al., 1991; Burr et al., 1994). For example, Ellsworth et al., (1991) noted initially an increase in vertical plume variance between soil surface and a depth of 2.5 m, followed by a decrease between 2.5 to 4 m, and an increase once again below 4 m depth. The longitudinal macrodispersion for the reactive solutes have been demonstrated to be much larger than that for nonreactive solutes, apparently caused by the variability of the sorption coefficient field (Burr et al., 1994).

3.2 *Chemical heterogeneity*

Whilst in recent years some studies examining the impact of physical heterogeneity (in the vertical direction) on solute transport have been carried out, the effect of chemical heterogeneity is relatively poorly understood. Bosma et al., (1993) considered the vertical heterogeneity in the vadose zone as it affects the sorption/retardation factor of a solute. Recently, Burr et al., (1994) examined the joint effect of heterogeneity of physical and chemical properties of an aquifer on field scale transport of both reactive and non-reactive solutes by groundwater. They showed that the joint effect of spatial variability in hydraulic conductivity and sorption coefficient can result in a pseudokinetic behaviour of large-scale sorption, leading to an increase in retardation factor with time and plume displacement distance, even when the underlying local scale sorption is assumed to be an equilibrium process.

The following examples show how the soil properties, which are known to vary (often markedly) with depth, affect the sorption behaviour of contaminants. Such chemical heterogeneities may have an important bearing on contaminant transport, as shown by Burr et al. (1994).

Sorption of toxic metals. Various soil properties (e.g. pH, OM content, chemical composition and ionic strength of soil solution, redox potential) control the sorption of different contaminants (e.g. heavy metals, nutrients, pesticides). Commonly in soils, pH varies significantly with depth, and can strongly influence the sorption properties of subsurface layers of soil for metals such as Cd, Pb, Cu and Zn. Most metal-substrate combinations exhibit rapid increase in metal retention over a narrow pH range, a phenomenon generally termed as adsorption edge. The sorption of metals in soils can indeed vary from 10% to 90% of applied concentrations over one or two pH units (Fig. 3). The pH at which the edge occurs tends to vary with the nature of the metal. Clearly, therefore, a small variation in pH near the metal sorption edge can have a major effect on its retention. Also, the presence of organic ligands and inorganic ionic species in soil solution, capable of forming stable complexes or competing for sorption sites, can significantly affect the retention of heavy metals by soils.

As discussed above, clay content in duplex soils increases sharply with depth. Both the content and type of clay minerals in soil affect sorption of heavy metals and may significantly affect their leaching behaviour. For example, in an agricultural soil in the vicinity of a lead smelter, Cd leached down though the light textured surface layer of soil to the top of a clayey B horizon in the profile (Cartright et al., 1977), but has remained immobilised in this layer for the last two decades (Fig. 4). This must have occurred due to higher sorption and lower permeability of the clayey layer which restricted any further downward movement of the contaminant. The variation

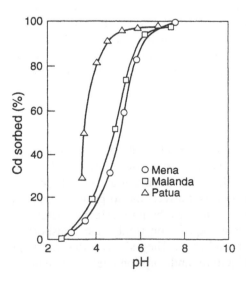

Figure 3. Cd sorption in soils as affected by pH (Naidu et al., 1994).

in Cd sorption with depth in this soil profile, presented in Table 1, explains the observed behaviour of Cd leaching (Fig. 4).

Role of DOC in metal sorption. The association of contaminants with naturally occurring organics can be a major factor controlling the fate and behaviour of contaminants in soils, surface water and sediments (Carter & Suffit, 1982; Dunnivant et al., 1992; Harter & Naidu, 1995). In addition to the potential for increased solubilities of the contaminants (Means et al., 1978; Enfield et al., 1989; Naidu et al., 1995), these organic phases (dissolved and particulate organics) can enhance the transport of contaminants through porous media. Such transport processes can occur either as soluble or particulate contaminant-organic complexes, such as stabilised alluvial colloid particles with adsorbed contaminants (Harter & Naidu, 1995).

Variations in concentrations of DOC with soil depth can complicate the interactions of contaminants in the soil environment because of the ability of DOC to form soluble complexes and chelates with the metals. Presence of natural organics in soil can lead to changes in both surface properties of soil and the nature (i.e. activity) of contaminants in the soil solution. Moreover, organics can induce anion exclusion through their effect on the density of surface negative charges of soil particles. Gruhn et al., (1985) used a combination of gel-chromatographic and molecular volume separation of dissolved organic substances to show different affinities of metals to low and high molecular weight organic substances. They reported that the concentrations of metals bound to high molecular weight organics decreased rapidly with depth. Metal-organic interaction is particularly important under Australian soil conditions in which the concentration of dissolved organics can be reasonably high in

Table 1. Variation of Cd sorption with depth in a contaminated soil (see Fig. 4).

Soil depth (cm)	Sorption coefficient (L kg^{-1})	pH*
0-10	17.9	5.65
10-15	64.0	5.93
15-20	163.8	6.41
20-25	254.8	6.66
25-30	856.7	7.70

* As measured during sorption (1:30) in 0.01M Ca[NO$_3$]$_2$.

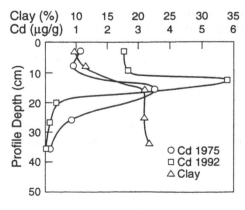

Figure 4. Cd leaching in a contaminated soil profile (Cartright et al., 1977).

the surface horizons of sodic soils due to the presence of high pH and concentrations of sodium in the soil solution.

At the soil pH range commonly found in Australian soils, many organic molecules are naturally negatively charged and can be sorbed on clays depending on their anion exchange capacities. In soils that have high anion exchange capacity, sorption of dissolved organics can lead to dispersion of fine colloids (Naidu et al., 1993). Such dispersive processes can enhance formation of immobile water domains, particularly in the surface horizons, as well as restrict the flow of water by clogging the micro-/ macropores.

The restriction in water movement often leads to the development of anoxic conditions in the surface and sub-surface horizons. Such reducing conditions can have a marked impact on the chemical behaviour of contaminants. Indeed, Naidu et al., (1992) have demonstrated in the field that reducing conditions in sodic soils can lead to mineral structure breakdown and lowering of pH, which have strong implications for reactions of phosphorus in soils.

Where the soils have dense, impenetrable sodic B horizons, the mobile colloids are laterally transported in subsurface waters into streams (Naidu et al., 1993). This lateral movement of colloids has been shown to be associated with phosphorus and metals such as Al and Fe (Naidu et al., 1995), indicating that simulations of contaminant transport using some existing models can over-emphasise the vertical movements of contaminants.

Sorption of pesticides. Soil pH also plays a major role in determining the sorption of pesticides in soils. A number of commonly used herbicides are ionogenic i.e. they transform from their molecular form to anions or cations, depending on their chemical nature. Weakly acidic herbicides (e.g. 2,4-D, sulfonylureas) upon dissociation become anions at pH higher than their pKa values (dissociation constant), whereas weakly basic herbicides (e.g. triazines) get protonated and become cations at pH lower than their pKa. The pH distribution in a soil profile will therefore determine dissociation or protonation of a compound and may greatly modify its sorption on the common negatively charged soils, thus affecting its leaching.

For sorption of non-ionic pesticides in soils, organic matter (OM) is considered as the most important soil constituent. OM contents, however, decline exponentially with depth in a soil profile. Figure 5 shows the variation of organic matter with depth in 57 profiles of agricultural soils in Montana, USA (Jersey & Nielsen, 1992).

A simple screening model developed by Jury et al., (1987) recognises this variation and includes a depth dependent rate of pesticide degradation. From a sorption point of view, however, the profile is assumed in the model to have uniform OM content. Recently, Kookana & Aylmore (1994) incorporated the exponentially declining OM content in a screening model and thereby considered depth dependence of both pesticide sorption and degradation in a soil profile. They showed that the leaching of pesticides is also markedly affected by the variations of sorption with depth in a soil profile (Table 2).

Degradation of pesticides. Chemical properties not only affect the sorption but also the degradation of pesticides. For example, degradation of a group of popular herbicides (sulfonylureas) has been shown to depend strongly on soil pH (Fig. 6). Sulfonylureas are known to undergo rapid hydrolysis in the acidic pH range. In acid soils the half-lives of these herbicides have been noted to be short, i.e. in months,

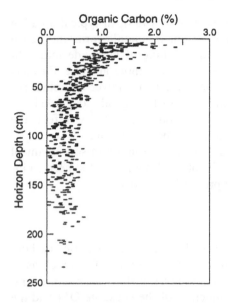

Figure 5. OM distribution in 57 soil profiles from Montana (after Jersey & Nielsen, 1992).

Table 2. Residual mass (%) for six pesticides based on measured sorption and degradation parameters at 1.0 m yr^{-1} recharge rate (source: Kookana & Aylmore, 1994).

Pesticide	Koc (L kg^{-1})	Half-life (days)	Residue, Constant OM	Residue, Variable OM
Fenamiphos	0.55	30	6.2×10^{-9}	7.2×10^{-5}
Linuron	0.84	143	0.06	5.0
Metribuzin	0.25	46	0.08	5.0
Metalaxyl	0.09	48	7.0	29
Prometryne	1.22	64	3.2×10^{-11}	6.5×10^{-5}
Simazine	0.15	75	6.7	31

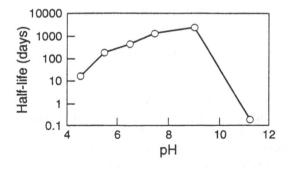

Figure 6. pH dependence of sulfonylurea hydrolysis (Blacklow & Pheloung, 1991).

whereas in alkaline soils (pH < 9) these herbicides can persist for years (Blacklow & Pheloung, 1991). The variation of soil pH with depth in a soil profile can have important consequences under field conditions. For example, the higher persistence of sulfonylureas in alkaline soils of southern Australia is causing some concerns with respect to carryover of their residue into following leguminous crops and pasture, which are extremely sensitive to these herbicides (Ferris, 1993). In typical alkaline soils of southern Australia, these herbicides are present in anionic form and hence not sorbed by soils with high CEC. This results in leaching of herbicide into sub-surface layers of soil profiles, which in some cases are even more alkaline. Furthermore, sub-surface soil layers have lower OM content and hence lower microbial activity. The herbicides are therefore neither chemically nor biologically degraded, resulting in longer persistence in such soils. This could potentially affect the sustainability of cereal-legume rotations practised in southern Australia.

This practical problem demonstrates the importance of variation in soil chemical properties in modifying the behaviour of contaminants and highlights the need to incorporate the chemical heterogeneity of the soil profile in a model.

3.3   *Biological heterogeneity*

Degradation of pesticides. Biological activity in soil affects the behaviour of both inorganic and organic contaminants. Interactions of nutrients (N and P) as well as of pesticides in soils are mediated by microbial processes of transformation and degradation. Microbial activity in a soil profile is a function of the substrate OM and the soil environmental conditions such as moisture and temperature. For many pesticides, biological degradation is the major pathway of their dissipation, which is affected by several soil properties including moisture content, temperature, pH, OM content, microbial activity, redox potential etc. All these properties are known to vary with depth in a soil profile. OM is a particularly important property to consider with respect to biological degradation. If other factors (e.g. temperature, moisture content) are constant, the microbial population density of a soil is directly related to the nature and content of OM present. As discussed earlier, however, it is well recognised that OM content in a soil profile rapidly decreases with depth. Concurrently, the microbial activity varies with depth, which governs pesticide degradation. Therefore, any realistic prediction of pesticide or other contaminant movement through soil is not possible without due regard to the biological heterogeneity in the vertical direction (Jury et al., 1987).

4   INTER-DEPENDENCE OF PHYSICAL, CHEMICAL AND BIOLOGICAL HETEROGENEITIES

As previously pointed out, physical heterogeneity has now been well recognised. However, soils are heterogeneous not only with respect to their physical properties such as hydraulic conductivity but also to their chemical and biological characteristics, leading to spatially variable sorption and degradation parameters. All three 'types' of heterogeneity can affect longitudinal dispersivity of a contaminant. So far, however, the variability of physical, chemical and biological properties in soils has not been considered together, probably because of the extreme complexity involved.

The recently published work by Bosma et al., (1993) has considered the physical and chemical heterogeneities in terms of lumped parameters, i.e. hydraulic conductivity and retardation factors, respectively. This work has shown that the nature of interdependence of physical and chemical heterogeneities can have a major impact on the longitudinal dispersivity of a solute. Bosma et al., (1993) found that the effects of physical and chemical heterogeneity can, in some cases, counteract each other on plume disperson. For example, in the case of equal physical and chemical heterogeneity, the plume spreading in the longitudinal direction was found to be relatively small. On the other hand, a combination of either low sorption and large conductivities, or vice-versa, caused large solute spreading. The work has demonstrated that the correlation between physical and chemical parameters cannot be ignored during the modelling of contaminant behaviour. Although not understood yet, the interaction amongst all physical, chemical and biological variabilities could be even more complex and may have even stronger implications.

5 CONCLUDING REMARKS

In this paper we have discussed the role of vertical soil heterogeneity in terms of several soil physical, chemical and biological properties known to affect the transport behaviour of contaminants in soils. Whilst in recent years extensive efforts have been directed towards the understanding of lateral heterogeneity, vertical heterogeneity is yet to receive adequate attention by hydrologists and soil scientists. Natural field soils are relatively more homogenous in the horizontal direction than vertical, due to the process of soil genesis. Ideally, to model real field situations, both vertical and horizontal heterogeneities of the soil physical, chemical and biological properties and their interactions have to be considered. This clearly spells a modeller's nightmare, but represents the only faithful approach.

Soil scientists involved with the understanding of various processes governing fate and behaviour of contaminants in soils can contribute by quantifying the key parameters and their functional dependence on relevant soil (heterogeneous) properties, in the form that these could be included in the transport models. The modellers on the other hand, can develop suitable approaches capable of integrating the spatial variability of biogeochemical properties at lateral and vertical scales. Obviously, there is no dearth of challenges for those modelling non-conservative contaminant behaviour in the soil vadose zone.

REFERENCES

Blacklow, W.M. & Pheloung, P.C. 1991. Sulfonylurea herbicides applied to acidic sandy soils: A bioassay for residues and factors affecting recoveries. *Aust. J. Agric. Res.* 42: 1205-1216.

Bosma, W.J.P., Bellin, A., van der Zee, S.E.A.T.M. & Rinaldo, A. 1993. Linear equilibrium adsorbing solute transport in physically and chemically heterogeneous porous formations 2. Numerical results. *Water Resour. Res.* 29: 4031-4043.

Bresler, E. & Dagan, G. 1981. Convective and pore scale dispersive solute transport in unsaturated heterogeneous fields. *Water Resour. Res.* 17: 1683-1693.

Burr, D.T., Sudicky, E.A. & Naff, R.L. 1994. Nonreactive and reactive solute transport in three-dimensional heterogeneous porous media: Mean displacement, plume spreading, and spreading. *Water Resour. Res.* 30: 791-815.

Butters, G.L., Jury, W.A. & Ernst, F.F. 1989. Field scale transport of bromide in an unsaturated soil, 1, experimental methodology and results. *Water Resour. Res.* 25: 1575-1581.

Carter, C.W. & Suffit, I.H. 1982. Binding of DDT to dissolved humic materials. *Environ. Sci. Technol.* 16: 735-740.

Cartwright, B., Merry, R.H. & Tiller, K.G. 1977. Heavy metal contamination of soils around a lead smelter at Port Pirie, South Australia. *Aust. J. Soil Res.* 15: 69-81.

Dunnivant F.M., Jardine P.M., Taylor D.L. & McCarthy, J.F. 1992. Cotransport of cadmium and hexachlorobiphenyl by dissolved organic carbon through columns containing aquifer material. *Environ. Sci. Technol.* 26: 360-368.

Ellsworth T.R. & Jury, W.A. 1991. A three dimensional field study of solute transport through unsaturated, layered, porous media 2. Characterisation of vertical dispersion. *Water Resour. Res.* 27: 967-981.

Ellsworth T.R., Jury, W.A., Ernst, F.F. & Shouse, P.J. 1991. A three dimensional field study of solute transport through unsaturated, layered porous media, 1. Methodology, mass recovery, and mean transport. *Water Resour. Res.* 27: 951-965.

Enfield, C.F., Bengtsson, G. & Lindquist, R. 1989. Influence of macromolecules on chemical transport. *Environ Sci. Technol.* 23: 1278-1286.

Ferris, I.G. 1993. A risk assessment of sulfonylurea herbicides leaching to groundwater. *AGSO J. Aust. Geol. & Geophy.* 14: 297-302.

Gruhn, A., Matthess, G., Pekdeger, A. & Scholtis, A. 1985. Die Rolle der gelosten organischem Substanz beim Transport von Schwermetallen in der ungesattingten Bodenzone. *Z. Dtsch. Geol. Ges.* 136: 417-427.

Helling, C.S. & Gish, T.J. 1986. Soil characteristics affecting pesticide movement into groundwater. In: Garner, W.Y., Honeycutt, R.C. & Nigg, H.N. (eds), *Evaluation of pesticides in ground water.* ACS Symp. Series 315, pp 14-38.

Jersey, J.K. & Nielsen, G.A. 1992. Montana soil pedon database reference manual. Mont. Agric. Exp. Stn. Rep. No. SR-043, Bozeman, MT.

Jury, W.A., Focth, D.D. & Farmer, W.J. 1987. Evaluation of pesticide groundwater pollution potential from standard indices of soil-chemical adsorption and biodegradation. *J. Environ. Qual.* 16: 422-428.

Jury, W.A. & Roth, K 1990. *Transfer Functions and Solute Movement through Soil.* Birkhauser Verlag, Basel.

Jury, W.A., Sposito, G. & White, R.E. 1986. A transfer function model of solute transport through soil, 1. Fundamental concepts. *Water Resour. Res.* 22: 243-247.

Harter R.D. & Naidu, R. 1995. Role of metal-organic complexation in metal sorption by soils. *Adv. Agron.* 55: 219-263.

Kookana, R.S. & Aylmore, L.A.G. 1994. Estimating the pollution potential of pesticides to groundwater. *Aust. J. Soil Res.* 32: 635-651.

Means, J.L., Crerar, D.A. & Dubuid, J.D. 1978. Migration of radioactive wastes: Radionuclide mobilization by complexing agents. *Science* 2000: 1477-1481.

Naidu, R., Bolan, N.S., Kookana, R.S. & Tiller, K.G. 1994. Ionic strength and pH effects on surface charge and Cd sorption characteristics of soils. *European J. Soil Sci.* 45: 419-429.

Naidu, R., Fitzpatrick, R.W. & Hudnall, W.H. 1992. Chemistry of saline sulfidic soils with altered water regime in the Mount Lofty Ranges, S.A. *Proc. Int. Symp. 'Strategies for Utilizing Salt Affected Lands'* pp 477-490.

Naidu, R., deLacy, N.J., Hollingsworth, I.D. & Fitzpatrick, R.W. 1995. Seasonal changes of iron and dissolved organic carbon concentrations in streams in the Warren catchment, South Australia. In: Naidu, R., Sumner & Rengasamy, P. (eds), *Australian Sodic Soils: Distribution, Properties and Management,* pp 185-190. CSIRO Publ. Melbourne.

Naidu, R., Williamson, D., Fitzpatrick, R.W. & Hollingsworth, I. 1993. Chemistry of throughflow water above clayey sodic B horizons: Implications to catchment management. *Aust. J. Expt. Agric.* 33: 239-244.

Northcote, K.H. & Skene, J.K.M. 1972. *Australian Soils with Saline and Sodic Properties.* Soil Publ. 27, CSIRO Publ., Melbourne, Australia.

Russo, D. 1991. Stochastic analysis of simulated vadose zone solute transport in a vertical cross section of heterogeneous soils during nonsteady water flow. *Water Resour. Res.* 27: 267-283.

Sudicky, E.A. 1986. A natural gradient experiment on solute transport in a sand aquifer: Spatial variability of hydraulic conductivity and its role in the dispersion process. *Water Resour. Res.* 22: 2069-2082.

Wagenet, R.J. & Rao, P.S.C. 1989. Modelling pesticides fate in soils. In: Cheng, H.H. (ed.), *Pesticides in the Soil Environment: Processes, Impacts and Modelling,* pp 51-78. Soil Sci. Soc. Am. Inc. Madison, WI.

CHAPTER 3

Solute transport in the surface soil

V.O. SNOW
CSIRO Division of Water Resources, Adelaide, Australia

B.E. CLOTHIER
HortResearch, Palmerston North, New Zealand

R.W. TILLMAN
Department of Soil Science, Massey University, New Zealand

ABSTRACT: Solute transport near the soil surface is complex and as yet poorly understood. This paper argues that a greater understanding of the processes influencing the movement of solutes into and out of the immobile water near the soil surface will enhance our ability to predict and control leaching of fertilisers and contaminants. These points are illustrated here via experiments and modelling. The paper demonstrates that the initial water content of the surface soil will greatly affect subsequent solute movement, particularly in the short term. Below the surface soil, solute transport can be predicted successfully. A new technique for independently assessing the immobile component of the soil water is described and then the paper examines how the immobile water content varies with soil water tension. Data showing invariance of the immobile water content over a wide range of time and length scales are then compiled. The paper concludes with a description of core leaching experiments designed to elucidate the movement of solute into the immobile water in response to wetting and drying of the soil surface.

1 INTRODUCTION

Those seeking to understand the movement of fertilisers and pesticides applied to the surface of field soils seem to fall into two camps. One group would argue that the environmental and soil conditions operating in the field cannot be duplicated either in the laboratory, or in controlled leaching experiments. In support of this belief they monitor the progress of chemicals applied to the soil surface in response to rainfall and evaporation without the intention of controlling soil or environmental conditions during leaching. While the chemicals are subject to a range of processes normally operating in the field, difficulty in accounting for extraneous influences precludes attempts to isolate, quantify, and understand these processes. Therefore the extrapolation of the results from these experiments, in either space or time, is fraught with difficulty.

An opposing point of view is that uncontrolled field experiments, although representing transport under natural conditions, do not allow those conditions to be measured or processes to be understood. Members of this group seek to control the local environment during solute transport, often using 'intact cores' in the laboratory, in order to understand better the processes affecting solute transport. However they often ignore processes operating in the field. Hence the applicability of their results to real-life conditions is just as questionable as those deriving from monitoring-style experimentation.

Here we present evidence that suggests that an understanding of water and solute movement in the top few centimetres of the soil, particularly soon after a chemical has been applied, will help close the gap between these two groups. Behaviour of solute in the near-surface soil is often the determinant of subsequent chemical mobility. Understanding this behaviour, combined with applying this understanding to both laboratory and field experimentation, will enhance our ability to predict the fate of chemicals in the field.

Movement of surface-applied chemicals into surface soil is a complex process that currently is not well understood. This is a serious omission as the early movement of fertilisers and contaminants can have great impact on their subsequent movement. This paper discusses the importance of the initial disposition of solute on subsequent leaching. It demonstrates that with sufficient prior knowledge of both the soil and the initial conditions of the solute and water it is possible to model solute movement. The paper also discusses a recent advance in the measurement of the partitioning of mobile and immobile water, and follows this by presenting a compilation of data demonstrating the invariant nature of the immobile water content. Finally, we conclude with the description of some preliminary experiments designed to elucidate, in the laboratory, the processes and rates of solute movement into the immobile soil water.

2 THE INITIAL DISPOSITION OF SOLUTE IN THE SURFACE SOIL

The immobile-mobile convection-dispersion equation (CDE) (van Genuchten & Wierenga, 1976) has proved to be a useful model with which to consider solute movement. In this model, pore water is divided in two classes. The pores containing water held between the air dry state and some critical water content, θ_c (m^3 m^{-3}), are considered to hold immobile water θ_{im} (m^3 m^{-3}). Additional water is considered mobile and denoted θ_m (m^3 m^{-3}). Solute may enter the immobile water by convection with invading water provided the soil's water content, θ, is less than θ_c, and can enter or leave θ_{im} in response to diffusion gradients. Solute within the mobile water will be subject to both convection and hydrodynamic dispersion. Mathematically, assuming fixed θ_c, this is,

$$\theta_m \frac{\partial C_m}{\partial t} + \theta_{im} \frac{\partial C_{im}}{\partial t} = \theta_m D_m \frac{\partial^2 C_m}{\partial z^2} - \theta_m v_m \frac{\partial C_m}{\partial z}$$

$$\theta_{im} \frac{\partial C_{im}}{\partial t} = \alpha \left(C_m - C_{im} \right)$$

(1)

where C_m and C_{im} are the mobile and immobile solute contents (g m^{-3}), t is time (day), z (m) is depth in the soil, D_m (m^2 day^{-1}) is the solute dispersion coefficient within the mobile water, v_m (m day^{-1}) is the pore water velocity, and α (day^{-1}) is the mass transfer parameter describing diffusion between the immobile and mobile phases.

Tillman et al. (1991) used this mobile-immobile CDE to investigate the effect of the antecedent water content on solute transport. They showed that the initial condition of the soil surface could substantially alter subsequent solute transport. In their experiment a bromide tracer was applied to the Manawatu fine sandy loam as a short pulse dissolved in 5 mm of water that was then irrigated, without ponding, onto the soil with a further 50 mm of tracer-free water. In one treatment the soil had been first pre-irrigated with 20 mm of water, while in the second treatment the solute was applied without pre-irrigation. The antecedent water content in the top 20 mm of soil was 0.13, and after irrigation rose to 0.4. Mean bromide concentrations in soil cores taken 24 hours after the start of irrigation are shown in Figure 1. Bromide applied to the drier soil was strongly retained in the top 60 mm of soil, while although bromide applied to the pre-wetted soil showed some retention close to the surface, there was also a definite peak in concentration much deeper at 250 mm. Also shown on Figure 1 are the results of the mobile-immobile CDE model simulations of Br distribution for the two regimes. The data and simulation agreed well. All parameters, except θ_c for which a value of 0.18 was found to fit the data well, were determined independently of the data.

This work demonstrated the utility of the mobile-immobile CDE for describing solute movement in the field, albeit on a small-scale under controlled conditions. Sensitivity analysis carried out by Tillman et al. (1991) indicated that θ_c and α were more important to determine accurately than the dispersion and pore-water velocity parameters that have traditionally been concentrated upon. These results also highlighted the importance of the timing of solute application in relation to rainfall or irrigation. Alternatively, where irrigation is available, it may be possible to control subsequent leaching of solutes by manipulating the initial water content of the soil surface prior to application of the chemical.

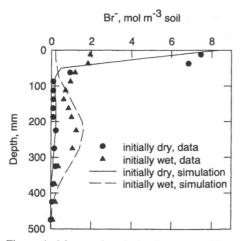

Figure 1. Measured and simulated bromide concentration profiles resulting from application to field-moist or pre-wetted soil (Tillman et al., 1991).

In another field study on the same soil, Snow et al. (1994) demonstrated that once below the soil surface, solute appeared to remain in the mobile water. Progress of the solute down the soil profile was quite predictable, despite the presence of soil horizons of contrasting texture with sharp interfaces. In this study, Snow et al. (1994) carried out a leaching experiment in a lysimeter of surface area 2 m^2 and 1 m deep. The soil consisted of 400 mm of fine sandy loam, underlain by 500 mm of fine sand and a further 100 mm of coarse sand. The lysimeter had been constructed from repacked Manawatu fine sandy loam, a recent alluvial soil, in 1967, some 23 years before the leaching experiment. Drainage from the lysimeter was collected under slight tension at a depth of 1000 mm. Soil solution samplers were installed at depths of 250 mm, 550, and 760 mm. Water flow was unsteady, but the data were analysed invoking steady-state assumptions similar to those in Roth et al. (1991) and Jury et al. (1982). The soil was pre-irrigated and allowed to drain overnight before a pulse application of KBr was applied to the soil surface as a solid. Intermittent irrigation followed. The data were normalised to a steady water flow of 10 mm day^{-1}.

The mobile-immobile CDE has four descriptive parameters, namely θ_c, D_m, v_m, and α. If no information is available a priori then a number of combinations of the parameters may be found to fit the data reasonably well, casting doubt on any mechanistic interpretation of these parameters. Where transfer between the mobile and immobile phases is relatively unimportant compared to convection and hydrodynamic dispersion $\alpha \cong 0$, and a simpler form of Equation (1), similar to the one-dimensional CDE may be employed. Here,

$$\frac{\partial C_m}{\partial t} = D_m \frac{\partial^2 C_m}{\partial z^2} - v_m \frac{\partial C_m}{\partial z} \qquad (2)$$

Any exchange of solute between the mobile and immobile water will appear as decreased pore-water velocity estimates and increased dispersion. If the water flux density, J_w (m^3 m^{-2} day^{-1}), is known, then the apparent transport porosity, θ_{st} (m^3 m^{-3}), can be calculated as J_w/v_m. Comparison of θ_{st} to the wetted soil water content, θ, will indicate the presence, or absence, of immobile water.

Equation (2) was applied to the data, of Snow et al. (1994) within a transfer function (Jury & Roth, 1990) framework. Briefly, the output solute concentration was expressed as the convolution integral of the solute concentration at the input depth, z_{in}, and the transfer function, f (day^{-1}),

$$C_m(z_{in} + z, t) = \int_0^t C_m(z_{in}, t') f(z_{in} + z, t - t') dt' \qquad (3)$$

where

$$f(z, t) = \frac{z}{\sqrt{4\pi D_m t^3}} \exp\left(-\frac{(z - v_m t)^2}{4 D_m t}\right) \qquad (4)$$

Firstly, the solute input at the soil surface was assumed to be a Dirac delta function, and D'_m and v_m were determined by fitting Equations (3) and (4) to the data. The results are shown in Table 1.

Table 1. CDE parameters, D_m (mm² day⁻¹) and v_m (mm day⁻¹), calibrated with either a surface Dirac delta function, or the data measured at 250 mm as the input function (Snow et al., 1994).

Output depth	Input, Dirac delta function at $z_{in} = 0$ mm		Input, data measured at $z_{in} = 250$ mm	
(mm)	D_m (mm² day⁻¹)	v_m (mm day⁻¹)	D_m (mm² day⁻¹)	v_m (mm day⁻¹)
250	220	21		
550	370	28	720	37
760	460	28	850	38
1000	470	31	630	36

Despite normalising the water flux to a constant value, it can be seen that fitted v_m increased with depth. As a result, θ_{st} decreased from 0.48 at 250 mm deep to 0.30 at 1000 mm. The water content of the soil was measured several times and found to be quite invariant, with θ at all depths being 0.36. Without invoking an assumption that there was some immobile water in this soil, there was no mechanism known to be operating which would explain the decrease in θ_{st} with depth. It is also only with the assumption that there is some immobile water retarding solute movement that we can explain why the *apparent* solute transport porosity, θ_{st}, is greater than θ and should not be regarded as equivalent to θ_m.

The pattern of a high value of θ_{st} near the soil surface followed by a decrease to more reasonable values has also been observed by Starr et al. (1986) and Butters et al. (1989), and is consistent with retention of the solute within immobile water near the soil surface but travelling in the mobile water once below the soil surface. Solute transport in the top few centimetres of the soil is troublesome to understand and therefore difficult to model. It is there that the water and solute fluxes are at their most transient and likely to deviate most from the assumptions. More importantly, although a pulse application of solid tracer was applied to the soil surface, it is not known if this resulted in a pulse application into the moving, or mobile soil-water.

To avoid this troublesome part of the soil, the CDE of Equation 2 was recalibrated. This time the input function was not a Dirac delta function as previously-assumed, but rather the input function was the data measured by the soil solution samplers at 250 mm. Now using the integral form of the transfer function equation, it was possible to obtain, by non-linear regression, D_m and v_m parameters to describe transport between the depths of 250 mm and 550 mm, 250 mm and 760 mm, and so on. These parameters are also shown in Table 1. Then the calibrated parameters remained constant with depth. Thus the depthwise variation in D_m and v_m previously obtained is actually considered to be an artefact of the incorrect representation of the surface conditions. The data, as well as the CDE calibrated between 250 mm and 550 mm, plus predictions of solute transport to 760 mm and 1000 mm are shown in Figure 2. It can be seen that excellent predictions of solute transport are obtained despite the assumptions simplifying the unsteady drainage and changes in soil texture and structure with depth. This suggests that if we can predict what will happen to the solute in the top few centimetres of soil, we might then have the ability to predict movement further down the soil profile.

With the surface soil excluded, θ_{st} reduces to a constant and more realistic value of 0.22, which might be considered equal to θ_m. Calculation of θ_{st} and θ_m by inverse

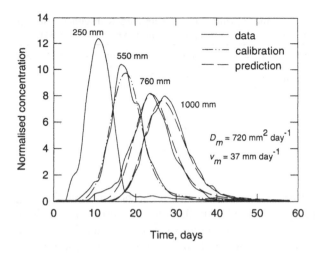

Figure 2. Data, calibration, and predictions of solute transport using the data measured at 250 mm as the input function (Snow et al., 1994).

methods from solute transport experiments would seem to be inherently risky as there is some danger that surface effects have not been taken into account.

3   RECENT ADVANCES IN THE MEASUREMENT OF θ_m

Clothier et al. (1992) recognised that direct methods for determining the amount of immobile water were required and they showed that a disk permeameter could be used for this purpose. The disk permeameter (Perroux & White, 1988) supplies solution to the soil surface at a slightly negative head. By filling the disk with tracer Clothier et al. (1992) proposed that it could be used to deduce the immobile water content of the soil. First a tracer-free solution is applied to the soil surface to pre-wet the soil so that all the water will be tracer-free initially. After infiltration reaches steady-state, the solution in the permeameter is replaced with a tracer solution of known concentration. Subsequent infiltration with tracer solution replaces only that water in the soil pores containing mobile water. As shown in Figure 3, by comparing the resident concentration in the soil beneath the disk with the concentration of the solution applied, θ_m can be determined. When this technique was applied to the surface soil of the Manawatu fine sandy loam, a θ_m of 0.21 was obtained. This compares favourably with the values of 0.18 from Tillman et al. (1991) and 0.22 from Snow (1992).

More recently Clothier et al. (1994; 1995a) have applied this disk permeameter technique to soils at different tensions in order to deduce the effect of the hydraulic regime on θ_m. The results of this are shown in Figure 4. Preliminary analysis suggests that at tensions of between −150 and −20 mm, θ_m is quite constant and that for modelling purposes the assumption of a constant value for θ_c would appear adequate. However more recent work suggests that although for a particular soil θ_m might be quite constant with changing soil water tension, it might vary greatly between soils (Clothier et al., 1995b).

Further work is aimed towards observing and understanding the role of soil pores

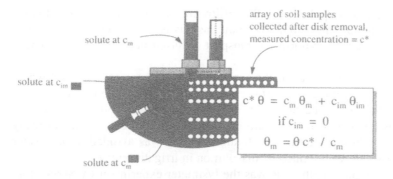

Figure 3. Measurement by a tracer-filled disk permeameter of the partitioning of the soil's water into the mobile and immobile fractions (Clothier et al., 1994).

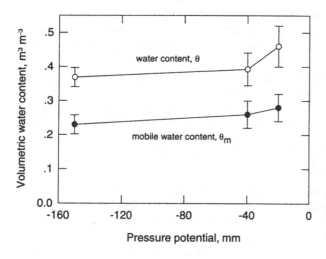

Figure 4 The effect of the disk pressure potential on soil water content and the mobile water content (Clothier et al., 1995a).

in water and solute transport. This should enhance our ability to predict and ultimately control solute movement in the field.

4   TIME AND LENGTH SCALE EFFECT ON θ_{im}

The preceding sections have presented data from three independent experiments done on the Manawatu fine sandy loam, albeit under quite different management and at varying time and length scales. It is illuminating to compare the estimates of θ_{im} resulting from these experiments.

At the shortest time and length scale was the disk permeameter experiment by Clothier et al. (1994) where solute infiltrated with about 100 mm of water, during a time span of a couple of hours. This experiment was undertaken within a herbicide strip of an apple orchard with the top 5 mm of the soil removed before the disk was applied. The water flow during solute movement was steady, at a constant potential of –20 mm, and solute was applied with the infiltrating water. For purposes of comparison, we will use the θ_{im} measured at –20 mm as this is likely to be the closest to the potential under which the majority of transport occurred in the other two experiments.

Next on the time and length scale is the experiment by Tillman et al. (1991) where solute applied to the pre-wetted soil surface penetrated to about 350 mm during the 24 hours between application and sampling. This experiment was done within a dairy farm where the pasture had been mown closely. The water application was unsteady and was applied only during the first few hours. Ponding was avoided during water application and the solute was applied by dissolution in irrigation water.

Largest on the time and length scale was the lysimeter experiment by Snow et al. (1994) where the progress of solute to 1000 mm was observed over a period of 2 months. Again the soil was a Manawatu fine sandy loam but the lysimeter had been repacked some 23 years previously. The lysimeter was managed as part of a dairy pasture. Solute was applied as a solid to a moist soil surface. Water flow was intermittent resulting from both irrigation and rainfall. Ponding was not observed during irrigation and was unlikely to have occurred during rainfall. For comparison with the other two experiments θ_{im} from the data measured at 250 mm and 550 mm will be used. θ for this layer was 0.39.

Table 2 shows θ_{im} from these three experiments. Despite the differences in time and length scale, soil management, and method of water and solute application, θ_{im} remains quite constant. Invariance of θ_{im} to the time and length scale might be expected given that Clothier et al. (1995a) showed that over a time of 2 to 14 days there was little diffusive exchange of solute between the immobile and mobile phases of the soil water. Although the experiment described by Snow et al. (1994) lasted for two months, solute concentrations were significantly elevated at any particular depth for only 10 to 20 days. Continual, albeit intermittent, additions of water during the 2 months ensured that θ did not fall below 0.18, the value of θ_c found by Tillman et al. (1990), so that any solute entering the immobile water could only have done so via diffusion. When concerned with solute transport under conditions where the soil water content does not fall below θ_c, for example during irrigation or perhaps winter when rainfall is comparatively high and evaporation low, a simplifying assumption of no exchange of solute between the mobile and immobile water may be adequate.

This again gives us optimism that effort put into discerning the process affecting solute movement will lead to greater understanding of, and ability to predict, solute movement in the field. For all these experiments it should be noted that water flow during solute movement was non-ponded. Therefore we still need to understand the movement of solute when macropores are water-filled and actively transporting solute.

Table 2. Immobile water content measured in the Manawatu fine sandy loam at varying time and length scale.

Source	θ_{im}	Time scale	Length scale
Clothier et al. (1994)	0.18	2 hours	100 mm
Tillman et al. (1991)	0.18	24 hours	350 mm
Snow et al. (1994)	0.17	2 months	1000 mm

5 MOVEMENT OF SOLUTE INTO THE IMMOBILE WATER

While there appears to be some progress in our ability to observe the partitioning of water in the surface soil into mobile and immobile components, what remains to be understood are the effects of different solute and water input regimes on that partitioning and subsequent transport. Before being leached, fertiliser applied to the soil surface will be subject to varying degrees of incorporation into the soil, dissolution, and convection, dispersion and diffusion into the immobile water. We need to understand these effects as well as to be able to predict transport of solutes entrapped in the immobile water. This section presents laboratory experiments where solute applied to the surface of cores of intact soil was subjected to various treatments before being leached.

Recently we leached three cores (100 mm in length and 100 mm in diameter) of an undisturbed soil. Various halide tracers were applied to the surface in order to investigate the effect of surface wetting and drying on subsequent mobility. First the cores were drained to 'field capacity' and allowed to dry for 24 hours before Br⁻ dissolved in water was applied to the soil surface. Evaporation dried the surface of the soil cores over the next 24 hours, before they were again wetted, this time with tracer-free water. On the third day the dried soil surface was irrigated with a solution containing Cl⁻. The following day the cores were leached under unsaturated water flow, with the upper and lower ends of the core maintained at 50 mm of tension. The

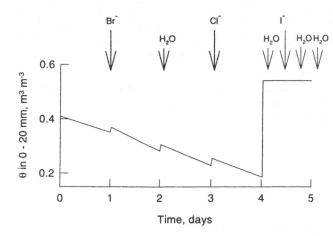

Figure 5. Diagram showing the variation in water content before final leaching of the cores. The time at which the tracers were applied is also shown.

Table 3. The percentage of tracer remaining in the top 20 mm of soil after 2.7 pore volumes of leaching. Both the mean and standard deviation are given. All means are significantly different at the 5% level.

Tracer	Tracer remaining (%) in the top 20 mm	
	Mean	St. dev.
Br⁻	17	2
Cl⁻	45	11
I⁻	7	2

first 25 ml of irrigation was with a tracer-free solution. This was followed with a further 25 ml of I⁻ tracer, while the remainder of the 235 ml irrigation applied to the soil surface was tracer-free. This scheme is shown in Figure 5. After 2.7 pore volumes of solution passed through the top 20 mm of soil, the core was sectioned and resident concentration of the tracers determined on duplicate samples from each core.

The percentages of applied Br⁻, Cl⁻, and I⁻ that remained in the top 20 mm of soil after 2.7 pore volumes of drainage are shown in Table 3. Despite the fact that the I⁻ tracer underwent less leaching (2.1 pore volumes) than the Br⁻ or Cl⁻ tracers (2.7 pore volumes), removal of I⁻ from the soil was considerably more complete.

When applied in the field, fertiliser will usually experience some degree of wetting and drying before a significant leaching event. Currently we do not understand enough of the effect of this wetting and drying on solute distribution. These effects will certainly be important for initial movement and leaching of solutes. Effects in the longer-term are less clear. Retention near the soil surface, in the more biologically-active soil, may increase degradation of a chemical, and management to facilitate movement of applied chemicals into the immobile water has the potential to reduce leaching to the groundwater.

6 CONCLUSIONS

Fertilisers and pesticides applied in the field are likely to undergo varying degrees of wetting and drying before being leached from the soil surface. The work described in this paper shows that transport parameters are dependent upon the hydraulic regime at the surface at the time of application and soon thereafter. Parameters determined empirically without regard to these processes will lack veracity; varying for the same field across years, or even rainfall events.

To make further progress we suggest that both controlled-input and natural-condition experiments need to continue, but greater attention to the detail of solute transport in the soil surface will be necessary. This will provide the link necessary to make our advances in understanding from controlled-input experimentation applicable to the prediction of solute transport under natural conditions in the field.

REFERENCES

Butters, G.L. & Jury, W.A. 1989. Field scale transport of bromide in an unsaturated soil. 2. Dispersion modelling. *Water Resources Res.* 25: 1583-1589.

Clothier, B.E., Green, S.R. & Magesan, G.N. 1994. Soil and plant factors that determine efficient use of irrigation water and act to minimise leaching losses. *Transactions 15th World Congress of Soil Science* Vol 2a: 41-47., Acapulco, Mexico, 10-16 July, 1994.

Clothier, B.E., Kirkham, M.B. & McLean, J.E. 1992. In situ measurements of the effective transport volume for solute moving through soil. *Soil Sci. Soc. Am. J.* 56: 733-736.

Clothier, B.E., Heng, L., Magesan, G.N. & Vogeler, I. 1995a. The measured mobile-water content of an unsaturated soil as a function of hydraulic regime. *Australian J. Soil Res.* 33: 397-414.

Clothier, B.E., Magesan, G.N., Heng, L. & Vogeler, I. 1995b. In situ measurement of the solute adsorption isotherm of a soil using a disc permeameter. For submission to *Water Resources Res.*

Jury, W.A. & Roth, K. 1990. *Transfer Functions and Solute Movement Through Soil: Theory and Applications.* Birkhauser, Basel, Switzerland.

Jury, W.A., Stolzy, L.H. & Shouse, P. 1982. A field test of the transfer function model for predicting solute transport. *Water Resources Res.* 18: 369-375.

Perroux, K.M. & White, I. 1988. Designs for disc permeameters. *Soil Sci. Soc. Am. J.* 52: 1205-1215.

Roth, K., Jury, W.A., Fluhler, H. & Attinger, W. 1991. Transport of chloride through an unsaturated field soil. *Water Resources Res.* 27: 2533-2541.

Snow, V.O. 1992. Solute transport in a layered field soil. Unpublished PhD thesis, Massey University, Palmerston North, New Zealand.

Snow, V.O., Clothier, B.E., Scotter, D.R. & White, R.E. 1994. Solute transport in a layered field soil: Experiments and modelling using the convection-dispersion approach. *Contaminant Hydrology* 16: 339-358.

Starr, J.L., Parlange, J-Y. & Frink, C.R. 1986. Water and chloride movement through a layered field soil. *Soil Sci. Soc. Am. J.* 50: 1284-1390.

Tillman, R.W., Scotter, D.R., Clothier, B.E. & White, R.E. 1991. Solute movement during intermittent water flow in a field soil and some implications for irrigation and fertiliser application. *Ag. Water Manage.* 20: 119-133.

Van Genuchten, M.Th. & Wierenga, P.J. 1976. Mass transfer in sorbing porous media. I. Analytical solutions. *Soil Sci. Soc. Am. Proc.* 40: 473-480.

CHAPTER 4

Laboratory and field application of a twin disc infiltrometer for measurement of soil hydraulic properties

K.R.J. SMETTEM
Dept. of Soil Science and Plant Nutrition, University of Western Australia, Nedlands, Australia

P.J. ROSS
CSIRO, Division of Soils, Davies Laboratory,Townsville, Australia

R. HAVERKAMP
IMG, Université Joseph Fourier, France

J.Y. PARLANGE
Department of Agricultural Engineering, Cornell University, Ithaca, N.Y. USA

P.J. GREGORY
Department of Soil Science, The University of Reading, Whiteknights, Reading, UK

ABSTRACT: A twin disc infiltrometer is employed to simultaneously measure one-dimensional and three-dimensional cumulative infiltration. The assumption that flow at early times is essentially one-dimensional can introduce error when estimating sorptivity, resulting in estimates that are greater than the true one-dimensional sorptivity. In contrast, sorptivity calculations using our recently developed three-dimensional infiltration equation give edge effect corrections that result in quite precise estimates of sorptivity in both laboratory and field tests. The estimates are shown to be consistent with measurements obtained from the buffered inner ring of the twin disc. The difference between three-dimensional and one-dimensional infiltration is used to calculate the value of the infiltration parameter, gamma, from the twin disc experiment in the laboratory. This parameter, together with the correct one-dimensional sorptivity, provide the information necessary to calculate the contribution of gravitational flow during three-dimensional infiltration. The laboratory test shows that when three-dimensional infiltration is dominated by capillary effects, as is often the case, estimation of hydraulic conductivity using quasi-steady-state solutions may not be possible. Results from the field test in a well aggregated topsoil illustrate a dominance of the gravity term and show a general application of the method for obtaining parameters that describe three-dimensional infiltration.

1 INTRODUCTION

The disc infiltrometer, described by Perroux & White (1988), is widely used to estimate the sorptivity and hydraulic conductivity of field soils. At early times, the rate of infiltration from a disc source at a negative hydraulic head h_o into a soil at a uni-

41

form initial negative hydraulic head h_n is dominated by the capillarity of the soil and is independent of the dimensionality of the flow (Philip, 1969; 1986):

$$\lim_{t \to 0} \left[\frac{Q(t)}{\pi r^2} \right] = \frac{1}{2} S_o t^{-1/2} \tag{1}$$

where $Q(t)$ is the flow rate from the disc source (mm^{3}h^{-1}); t is time (h); $S_o = S(h_o, h_n)$ is the sorptivity (mm h$^{-1/2}$) and r is the disc radius (mm). Integration of Equation (1) with respect to t gives:

$$I = S_o t^{1/2} \tag{2}$$

where I is the cumulative infiltration (mm). In practice, the sorptivity is generally obtained as the slope of the I versus $t^{1/2}$ plot at 'early-times' using Equation (2).

For some soils and disc radii, geometry can intrude on the initial one-dimensional character of the absorptive flow at very short times (Philip, 1986), thereby introducing possible observational errors into estimates of sorptivity obtained using Equation (2). The time, t_{geom}, at which geometry can be expected to swamp the initially one-dimensional character of the process is (Philip, 1969):

$$t_{\text{geom}} = \left[\frac{r(\theta_o - \theta_n)}{S_o} \right]^2 \tag{3}$$

where θ_o is the volumetric water content at h_o (m^{3}m^{-3}) and θ_n is the volumetric water content at h_n.

In practice, t_{geom} cannot be evaluated a priori so that any error in estimating sorptivity from Equation (2) will lead to errors in t_{geom}. Cook and Broeren (1994) have also shown that infiltration data depart from linearity much earlier than given by t_{geom}. Warrick (1992) showed that for two soils with t_{geom} of < 1h and 30h, one-dimensional behaviour was only dominant for < 20s and 100s respectively.

Haverkamp et al. (1994) concluded that the determination of sorptivity from Equation (2) was questionable because of the very short times over which it could be applied.

Furthermore, errors in the estimation of sorptivity may then be carried over into estimates of hydraulic conductivity based on Wooding's (1968) steady-state solution, which is commonly expressed as (White et al., 1992):

$$K_o = \frac{Q_\infty}{\pi r^2} - \frac{4bS_o^2}{(\theta_o - \theta_n)\pi r} \tag{4}$$

where K_o is the hydraulic conductivity (mm h^{-1}) and b ($1/2 \leq b \leq \pi/4$) is a function of the shape of the soil water diffusivity function.

An alternative method evaluates K_o from measurements of S_o at two different supply potentials (White & Perroux, 1992). Again, Equation (2) is usually used to obtain S_o by this method.

Warrick (1992) refined the short-time solution by adding the geometric effect of the disc source to the one-dimensional solution. His analysis is partially empirical and assumes a constant average diffusivity $D = Kdh/d\theta$ (mm^2 h^{-1}), which is appropriate for the 'linear' soil model (Philip, 1973). This model defines one bound to the envelope of possible soil wetting patterns. The other bound is given by the sharp Green-Ampt type model, which assumes a Dirac δ function in which $D(\theta) = 0$, except at θ_o, where $D \to \infty$.

Based initially on this 'sharp-front' model, Smettem et al. (1994) and Haverkamp et al. (1994) have introduced and refined approximate solutions for flow out of an unconfined disc infiltrometer. Their simplified, explicit, one-dimensional and three-dimensional infiltration equations, valid over most of the experimental time range, are:

$$I_{1-d} = S_o t^{1/2} + \left[K_n + \frac{(2-\beta)}{3}(K_o - K_n) \right] t \tag{5}$$

and

$$I_{3-d} = S_o t^{1/2} + \left[K_n + \frac{(2-\beta)}{3}(K_o - K_n) + \frac{\gamma S_o^2}{r(\theta_o - \theta_n)} \right] t \tag{6}$$

where K_n is the hydraulic conductivity corresponding to h_n (usually assumed to be negligible compared to K_o), γ is a proportionality coefficient with bounds $0.6 < \gamma < 0.8$ (Haverkamp et al., 1994) and β is an integral shape constant introduced through the expression (Haverkamp et al., 1990):

$$\frac{K - K_n}{K_o - K_n} = \frac{\theta - \theta_n}{\theta_o - \theta_n} \left[1 - \frac{2\beta(\theta_o - \theta_n)}{\hat{S}_o^2} \int_{\theta_n}^{\theta_o} D(\bar{\theta})d\bar{\theta} \right] \tag{7}$$

where $\bar{\theta}$ is a variable of integration and $\hat{S}_o$ is an approximation to the sorptivity given by

$$\hat{S}_o^2 = 2(\theta_o - \theta_n)\int_{\theta_n}^{\theta_o} D(\theta)d\theta \tag{8}$$

For times over which gravity effects are negligible, Equation (6) reduces to (Smettem et al., 1995)

$$I_{3-d} = S_o t^{1/2} + \frac{\gamma S_o^2}{r(\theta_o - \theta_n)}t \tag{9}$$

This can be compared to the short time solution given by Chu et al. (1975) which is relevant for two-dimensional edge effects in linear heat diffusion. Using a constant diffusion coefficient $D^* = \pi S_o^2/4(\theta_o - \theta_n)^2$, an equivalent flux expression was developed by Warrick (1992)

$$\frac{Q}{\pi r^2} \cong 0.5\, S_o t^{-1/2} + 0.885\, S_o (D^*)^{1/2}/r \tag{10}$$

The integration of this expression with respect to time leads to

$$I = S_o t^{1/2} + \frac{0.885\, b^{1/2} S_o^2 t}{r(\theta_o - \theta_n)^2} \tag{11}$$

Comparing this expression with Equation (9) gives a value of $\gamma = 0.784$, which is within the bounds given by Haverkamp et al. (1994) and similar to the optimal value of 0.75 obtained from the experimental results of Smettem et al. (1994).

Obviously, application of Equation (2) or Equation (9) to the same set of data points will give different values of sorptivity. Cook & Broeren (1994) have recommended that it is prudent to obtain estimates of S and K using all available methods of analysis. An attractive alternative approach is to ensure that in addition to three-dimensional infiltration, the one-dimensional sorptivity is measured accurately in situ at the same location.

The twin disc infiltrometer described by Smettem et al. (1995) is a refinement of the single disc tension infiltrometer and may be used to obtain estimates of both one- and three-dimensional cumulative infiltration for the same initial and boundary conditions. One-dimensional infiltration data from the buffered inner disc permits evaluation of the sorptivity using Equation (2). The combined efflux from the inner and outer discs may be analysed for geometry effects using Equation (9) at early times and checked for gravity effects at longer times using Equation (6).

In this paper we present both laboratory and field data to illustrate the utility of the twin disc method for infiltration parameter identification.

2   METHODS

2.1   *Infiltrometer design*

Our design modifies the single disc tension infiltrometer of Perroux & White (1988). The design, shown in Figure 1, depicts a double ring quadrant 1/4 disc which was used in the laboratory experiments reported here. It comprises an inner disc surrounded by an outer 'buffer' ring. The two discs form part of the same basal unit but are not hydraulically connected. Water is supplied to both discs from the two calibrated water reservoirs. A single bubbling tower is used to control and balance the negative pressure head at the supply surface, h_o (mm), over the two discs. As in the Perroux & White (1992) design, h_o is given by

$$h_o = z_2 - z_1 \tag{12}$$

2.2   *Laboratory tests*

Laboratory tests were performed using square perspex boxes with sides of 0.5 m, packed with soil to a depth of 0.4 m. Because the flow field is radially symmetrical, a double ring quadrant 1/4 disc (Fig. 1) was used to supply water at a corner of the box. The combined flow from the inner and outer discs creates the full three-dimensional flow field over a radius of 150 mm. The radius of the buffered inner disc is 100 mm.

Experiments were performed at a negative pressure head of 30 mm, on a sandy loam described by Smettem et al. (1994). θ_n was 0.04 m^3m^{-3} and θ_o was 0.32 m^3m^{-3}. No contact material was necessary for these experiments.

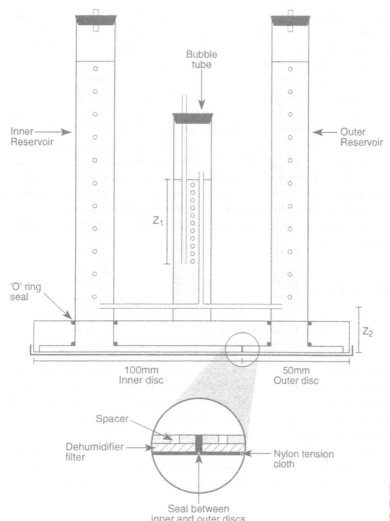

Figure 1. Diagram of the double ring infiltrometer.

2.3 *Field tests*

Field tests were performed on a well aggregated calcareous loam under permanent pasture at the Bridget's Research Station, Wellesbourne, Hampshire, UK.

The twin disc permeameter used for the field test had an inner disc radius of 62.5 mm and a combined inner and outer disc radius of 126 mm. The negative pressure head at the supply surface was 20 mm and measurements were made at a depth of 50 mm after removal of the surface pasture cover. Measurements were also made with a conventional single disc permeameter of 42.5 mm radius, again using a negative pressure head of 20 mm at the supply surface. A 5 mm thick capping of fine sand was used to establish good contact between the disc and the soil surface. The initial water content of the soil was measured gravimetrically by the weight loss from oven drying at 105°C. The water content at the supply surface was obtained at the cessation of measurement using the twin disc. The disc was carefully removed from the soil surface, the sand was scraped away and a sample of soil about 5 mm thick was collected for water content analysis. The gravimetric water contents were converted to volumetric using bulk density values obtained from soil cores.

Four replicate measurements were made at the same time in June 1996 using small disc permeameters on a level area of 1.5 m². The measurements were performed with sufficient separation distance to ensure that the expanding radial wetting fronts did not coalesce. The separation distance was also sufficient to allow space for the twin disc to be placed on the soil surface without overlapping any areas that had been sampled for water content. At the completion of these measurements the sand capping was carefully removed and the soil surface was exposed to summer evaporative conditions for ten days. The twin disc test was then performed at the same location. Gravimetric sampling immediatly prior to performing the twin disc test revealed that the soil water content had returned to the value that was observed prior to performing the single disc tests.

3 RESULTS AND DISCUSSION

3.1 *Estimation of sorptivity from the laboratory test*

Figure 2 compares cumulative three-dimensional infiltration with cumulative flow from the buffered inner disc for three replicate laboratory experiments, plotted against the square-root of time. Also shown are data from independent measures of one-dimensional sorptivity obtained from soil cores packed in the same manner and to the same bulk density as the perspex box. It is evident that the slope of the 'early-time' three-dimensional infiltration (measured over the first 0.17 h) is greater than the slope of the cumulative infiltration within the buffered inner disc. The slope of the latter gives a sorptivity of 51 mm h$^{-1/2}$, which is close to the one-dimensional sorptivity of 50 mm h$^{-1/2}$, obtained from the soil cores. In contrast, the 'sorptivity'

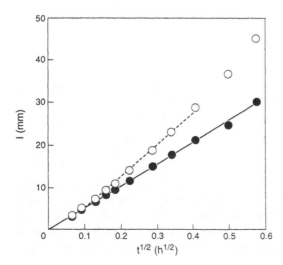

Figure 2. Cumulative infiltration for the laboratory experiment: Combined inner and outer discs (o), 'buffered' inner disc (●), 1-d flow in soil cores — and ···· regression of the 'early time' flow from the combined inner and outer discs.

identified from the cumulative three-dimensional flow gave a value of 72.5 mm h$^{-1/2}$, which is 45% higher than the one-dimensional measurement.The severe overestimate of sorptivity using Equation (2) would lead to inaccurate estimation of hydraulic conductivity, if used to solve Equation (4) for K_o.

3.2 *Estimation of hydraulic conductivity for the laboratory test.*

Apart from experimental errors, the errors involved in estimating the hydraulic conductivity from Equation (4) are firstly (as we have shown) overestimation of sorptivity using Equation (2) for cumulative 3-d infiltration at 'early-times' without correcting for the edge effect and secondly, the possibility that $Q/\pi r^2$ has not reached 'steady-state' during the experiment. Figure 3 shows an example of the components of 3-d flux at long times, as estimated using the accurate implicit equation of Haverkamp et al. (1994). The term t_{grav} in Figure 3 is the characteristic time at which gravity becomes significant (Philip, 1969), defined by

$$t_{grav} = S_o^2 / K_o^2 \qquad (13)$$

For this soil the one-dimensional hydraulic conductivity at this supply potential, obtained from permeameter measurements, is 9 mm h^{-1}. t_{grav} is therefore 25 h, a value considerably in excess of the experimental time period of 0.33 h.

To illustrate the possible errors that can occur in estimation of hydraulic conductivity using Equation (4), we use the laboratory experimental data reported here, taking $b = 0.55$ and $Q/\pi r^2 = 98.6$ mm h^{-1}, its value at 0.33 h, obtained from regression over the last four points of the experimental I_{3-d} data. The correlation coefficient of these last four points (r^2) is 0.999, yet the experiment is far from t_{grav}. Inserting these values into Equation (2) gives a highly inaccurate estimate of 55 mm h^{-1} for

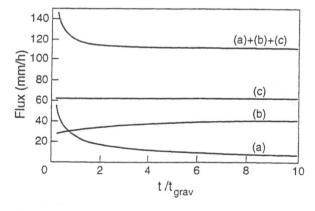

Figure 3. Approach to steady-state of the three-dimensional flux, illustrating contributions of a) the sorptivity component $S_o/2t^{1/2}$, b) the gravity component $(2 - b)K_o/3$ decreasing towards K_o, c) the 3-d edge effect $\gamma S_o^2/r(\theta_o - \theta_n)$. In this example the sorptive flux is balanced by the increase in the gravitational flux to produce an early approach to steady-state. Values used are $S_o = 50$ mm h$^{-1/2}$, $K_o = 50$ mm h^{-1}, $b = 0.5$, $\gamma = 0.7$, $r = 100$ mm and $(\theta_o - \theta_n) = 0.28$.

K_o. Note that this error cancels somewhat if the erroneous sorptivity of 72.5 mm $h^{-1/2}$ is used, resulting in a plausible estimate of 11 mm h^{-1}. Clearly, little faith can be placed in such a result.

This example illustrates that incorrect estimation of sorptivity has a large influence on the hydraulic conductivity calculated from Equation (4). The magnitude of the error in the calculated hydraulic conductivity is proportional to the error in the square of the sorptivity. It is also evident that use of Wooding's Equation (4) is not justified in this case because at 0.33 h the experiment has not yet reached 'steady state', despite the fact that the linear regression of the last four points gives such a good correlation coefficient.

3.3 *Field measurements*

One-dimensional infiltration from the buffered inner disc at the Bridget's site is shown in Figure 4. Cumulative infiltration was described using Equation (5), with K_n assumed to be negligible. Because β is unknown, but varies between 0 and 1 (Haverkamp et al., 1994), the gravity term $(2 - \beta)/(3 \cdot K_o)t$ gives bounds for K_o but cannot be used to obtain K_o exactly. The best fit to the data gives $S_o = 12$ mm $h^{-1/2}$ and a gravity term 'A'$=(2 - \beta)/(3 \cdot K_o)t$ of 25 mm h^{-1}. In contrast to the laboratory test, infiltration in this soil is influenced by a substantial gravity term that befits its macroporous character.

The cumulative three-dimensional infiltration from the combined efflux from the inner and outer discs was estimated from Equation (6) using the parameters obtained from the inner disc infiltration analysis, together with the measured water content difference, $\theta_o - \theta_n$, $= 0.2$ and assuming $\gamma = 0.75$. Results are shown in Figure 5 and give a very good representation of the measured three-dimensional infiltration.

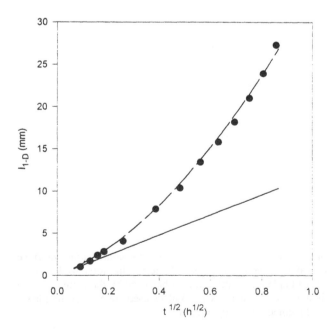

Figure 4. One-dimensional cumulative infiltration from the buffered inner disc at the Bridget's field site plotted against the square-root of time (●). Lines are Equation (5) fitted – – – and Equation (2) —.

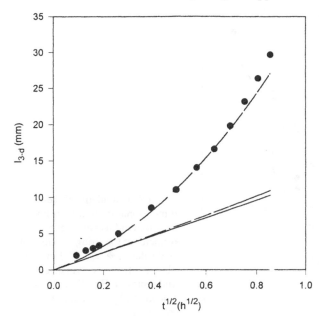

Figure 5. Measured cumulative three-dimensional infiltration from the twin disc infiltrometer at the Bridget's site (●). Lines are Equation (6) fitted (– – –), Equation (2) —. and Equation (9) – - –.

It is evident from Figure 5 that for this large diameter disc the edge effects are quite small. To provide a further test of the model we therefore used the parameters obtained from the twin disc experiment to describe cumulative three-dimensional infiltration from the smaller 42.5 mm diameter permeameter. Results shown in Figure 6 illustrate that although there is some variability between the four replicate small disc permeameter measurements, the average cumulative infiltration data is well described by the three-dimensional infiltration equation with parameters obtained from the twin disc experiment. It is also clearly evident that sorptivity would again be overestimated if the 'early-time' assumption of one-dimensional flow was applied to the analysis of this data.

3.4 *Practical time limits for the disc infiltrometer*

Results presented here bring us to the the consideration of practical experimental time limits. Although it is possible to examine I_{3-d} to check that it is near steady-state, there is a practical time limit related to the model assumption of a uniform initial profile. It is not realistic to continue an infiltration experiment beyond a profile depth of say, $y_1 = 150$ mm. Using for the sake of simplicity the non-gravity corrected expression of y_1 as given by Equation (12) of the paper by Smettem et al. (1994), an estimate of the experimental time limit is obtained from

$$t_{\exp} = \frac{\gamma_1^2 (\theta_o - \theta_n)^2}{S_o^2} \tag{14}$$

For the laboratory experiment $t_{\exp} = 0.66$h, which is twice the value of our experimental time but still is far smaller than t_{grav}.

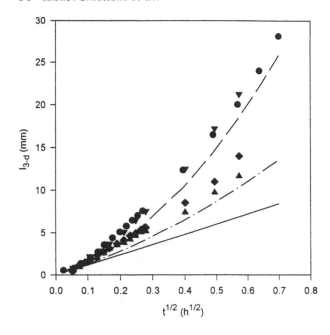

Figure 6. Cumulative three-dimensional infiltration from the 42.5 mm radius single disc infiltrometer at the Bridget's site (four replicates ●, ◆, ▲, ▼). Lines are obtained from the twin disc data and represent Equation (2) —, Equation (9) – - – and Equation (6) – – –.

For such times, K_o can be calculated from Equation (4) if an estimate of β appropriate for the soil is available, but errors may be large because of the large and uncertain values of the terms involving the sorptivity. In general, the use of the Wooding solution, Equation (2), for the calculation of hydraulic conductivity values K_o, from a disc infiltrometer experiment is questionable, because the method gives values which a priori cannot be identified with the 'real' hydraulic conductivity values. Whatever method is used, it is useful to calculate t_{geom} and t_{grav} to ascertain where on the time scale the experiment lies and whether the assumptions made are justified.

4 CONCLUSIONS

The double ring disc infiltrometer provides an accurate estimate of 1-d sorptivity and can also be used to measure three-dimensional infiltration. The analysis illustrates that the assumption of 1-d flow at 'early times' for the single disc infiltrometer may lead to erroneous estimates of sorptivity, with this error carrying over into the estimation of the hydraulic conductivity.

It is also shown that if cumulative 3-d infiltration is dominated by capillary flow effects, it is erroneous to attempt an estimation of hydraulic conductivity.

ACKNOWLEDGMENTS

The senior author acknowledges an Australian Academy of Science and British Royal Society exchange program grant. Dr D.E. Smiles and Prof. D.A. Barry are thanked for their comments.

REFERENCES

Cook, F.J. & Broeren, A. 1994. Six methods for determining sorptivity and hydraulic conductivity with disc permeameters. *Soil Sci.* 157: 2-11.

Haverkamp, R., Parlange, J-Y, Starr, L., Schmitz, G. & Fuentes, C. 1990. Infiltration under ponded conditions, 3, A predictive equation based on physical parameters. *Soil Sci.* 149: 292-300.

Haverkamp, R., Ross, P.J., Smettem, K.R.J. & Parlange, J.Y. 1994. Three-dimensional analysis of infiltration from the disc infiltrometer, II, Physically-based infiltration equation. *Water Resour. Res.* 30: 2930-2934.

Perroux, K.M. & White, I. 1992. Designs for disc permeameters. *Soil Sci. Soc. Amer. J.* 52: 1205-1215.

Philip, J.R. 1967. On solving the unsaturated flow equation, 1, The flux-concentration relation. *Soil Sci.* 116: 328-335.

Philip, J.R. 1969. Theory of infiltration. *Adv. Hydrosci.* 5: 215-305.

Smettem, K.R.J., Parlange, J.Y., Ross, P.J. & Haverkamp, R. 1994. Three-dimensional analysis of infiltration from the disc infiltrometer: A capillary-based theory. *Water Resour. Res.* 30: 2925-2929.

Smettem, K.R.J., Ross, P.J., Haverkamp, R. & Parlange, J.-Y. 1995. Three-dimensional infiltration from the disc infiltrometer, III, Parameter estimation using a double disc tension infiltrometer. *Water Resour. Res.* 31: 2491-2496.

Warrick, A.W. 1992. Models for disc infiltrometers. *Water Resour. Res.* 28: 1319-1327.

White, I., Sully, M.J. & Perroux, K.M. 1992. Measurement of surface-soil hydraulic properties: Disk permeameters, tension infiltrometers and other techniques. In: Topp G.C. et al. (ed.), *Advances in Measurement of Soil Physical Properties: Bringing Theory into Practice*. pp 69-103. SSSA Spec. Publ. 30, Soil Sci. Soc. Amer., Madison W.I., USA.

White, I. & Perroux, K.M. 1989. Estimation of unsaturated hydraulic conductivity from field sorptivity measurements. *Soil Sci. Soc. Amer. J.* 53: 1093-1101.

Wooding, R.A. 1968. Steady infiltration from a circular pond. *Water Resour. Res.* 4: 1259-1273.

CHAPTER 5

Seasonal dynamics of stream networks in shallow groundwater systems: A simple analytical approach

J.J. DE VRIES
Faculty of Earth Sciences, Vrije Universiteit, Amsterdam, Netherlands.

ABSTRACT: Surface water and groundwater are normally closely connected in areas with shallow aquifer systems. Stream systems can thus be considered as the outcrops of associated groundwater flows in areas with a shallow groundwater table and a pervious subsurface where almost all rainfall surplus infiltrates. The stream network must have the capacity to release the seasonally dependent precipitation surplus through the continuum of ground and surface waters. A stream network therefore consists of a hierarchical system of discharge-contributing branches of different order and incision depth, that contracts and expands with the seasonal fluctuation in recharge and water table depth. Coupling the mathematical expressions for groundwater drainage and stream flow enables development of a conjunctive model which relates stream density and channel size to the seasonal rainfall character for given geological and geomorphological conditions. This model further allows for assessment of a drainage network response to a changing environment.

1 INTRODUCTION

Surface water and groundwater are normally closely connected in shallow aquifer systems, notably in pervious sandy lowland areas where streams and lakes can be considered as the outcrops of shallow groundwater tables. The present study refers to shallow sandy aquifer conditions where almost all precipitation excess percolates to the subsurface to become part of a groundwater flow system. Under such conditions the groundwater receiving stream system can be considered as the interface between groundwater flow systems and the associated surface drainage network.

This interface must be able to discharge the precipitation surplus as groundwater and as stream flow. The size, the incision depth and the spacing of the stream are the geometric properties which determine its groundwater drainage capacity as well as its stream flow capacity in a given geological and geomorphological situation. Seasonal imbalance between recharge and discharge are adjusted by a fluctuating groundwater table.

This study proposes a deterministic approach to relate the geometric properties of the stream network for given geological conditions to the rainfall intensity and duration characteristics. The relations are developed in a theoretical model which is obtained by coupling the mathematical expressions for groundwater drainage, stream flow and groundwater level fluctuations for this continuous system of groundwater and surface water.

In previous studies the present author (De Vries, 1974, 1994) has shown that a stream system under the above mentioned conditions develops in equilibrium with and in response to a threshold value in rainfall and the accordingly required discharge. The stream spacing and channel dimensions were determined as dependent variables from mathematical flow relations with subsurface permeability, rainfall characteristics and large-scale topography as independent variables. In this way, first order stream systems have been synthesized as a function of a low frequency threshold value of rainfall and topography-related average groundwater depths. The strong relation between groundwater depth and stream spacing is illustrated with the observations depicted in Figure 1. This further shows a hierarchical stream network that contracts and expands with seasonal recharge variations and associated groundwater level fluctuations.

The present study investigates the difference in stream properties between the first order and second order stream systems as a function of their respective threshold rainfall-frequencies and their incision depths. The difference in channel geometry refers to Horton's bifurcation ratio.

2   MATHEMATICAL MODEL

2.1   *Reference area*

The conditions in the Netherlands Pleistocene sandy flatland will be used henceforth as a reference (Fig. 1). This region can be characterized as a Pleistocene fluvial fan with an average thickness of the upper aquifer of 50 m. The predominantly medium-grained deposits are covered by fine-grained fluvio-eolian sands (so called Coversands), with a thickness of up to 10 m. The whole forms an undulating topography with height differences of up to 3 m, whereas the average regional slope in the area is about 1:2500 (Fig. 1).

The climate is semi-humid with moderate temperatures, an average annual precipitation of 775 mm and an actual evapotranspiration of approximately 500 mm. During the summer half-year the precipitation and evaporation more or less counterbalance, so that the rainfall surplus of about 275 mm prevails during the winter period. Infiltration capacity is rarely exceeded because of the favorable entry characteristics of the permeable sandy soil, the flat topography and the moderate rainfall intensities.

The natural drainage system is characterized by a network of small streams. The spacing of the first order streams ranges from about 200 m for areas with a shallow groundwater table (< 0.5 m), to about 2000 m for areas with deeper water tables (> 2 m) (Fig. 1). Stream gradients vary from 1:500 for the steepest areas to 1:2500 for the most level areas (De Vries, 1974; Ernst, 1978).

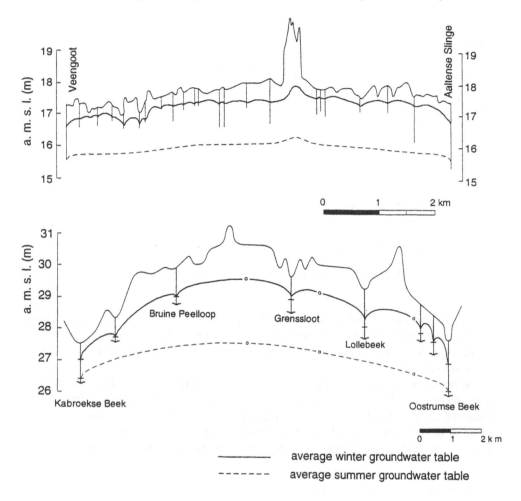

Figure 1 Hydrologic-topographic sections through some small stream systems in the Netherlands Pleistocene area, showing the relationship between groundwater depth and the seasonally expanding and contracting stream spacing. The lowest-order streams are mainly formed by a parallel network of ditches.

2.2 *Conceptual model*

The development of the mathematical expressions follows the following reasoning, assuming that all precipitation excess infiltrates to the subsurface (Fig. 2).

– The required discharge depends on the rainfall characteristics, notably the intensity-duration relation, and the available groundwater storage capacity.

– The groundwater drainage capacity of the stream system must be sufficient to cope with the required discharge. The stream characteristics that determine the drainage capacity include stream spacing, channel size and maximum hydraulic gradient.

– Stream flow capacity of all stream orders must be able to receive and transport groundwater discharge from the first order drainage system. Stream flow is controlled by channel geometry, channel roughness and stream bed gradient.

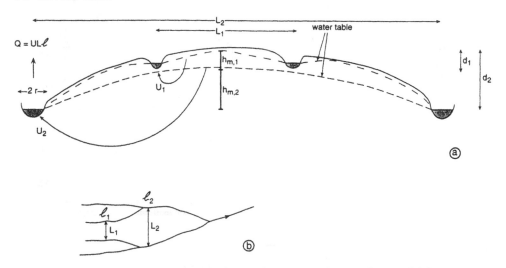

Figure 2 Conceptualization of the combined groundwater (a) and stream flow model (b).

– Stream systems of different order will respond to different rainfall characteristics which can be chacterized by their frequency of occurrence. Higher rainfall amounts occurring with lower frequency will activate lower order streams through a rise of the groundwater table.

The actual processes are complicated and of a non-linear nature. The present theoretical model is only a conceptual approximation of reality, intended to elucidate the functional relationship between the geological/geomorphological and climatic properties of an area on one side and the characteristics and behavior of the associated drainage network on the other. Predictive adequacy is not the main purpose of this model.

2.3 *Required discharge capacity*

The discharge capacity required to cope with precipitation excess depends on the combination of rainfall characteristics and storage capacity. In order to avoid exceeding groundwater storage capacity, which would result in ponding and marshy conditions, the following condition has to be satisfied:

$$t\overline{U} \geq ti - S \quad [L] \tag{1}$$

where t is period length of rainfall, i is average rainfall rate per time unit, $\overline{U}$ is average discharge, S is the available groundwater storage capacity in the unsaturated zone.

The total storage capacity between depth d and 0 is denoted $S(d)$. The relation between the mean rainfall amount per time unit that is exceeded with a given probability in some time interval and the length of that time interval, can be expressed by a rainfall intensity-duration-frequency relationship of the type:

$$i = ct^{-m} \tag{2}$$

where i is the mean amount of precipitation per time unit that is exceeded with a given probability p in a time interval t; c and m are local climatic parameters depending only on p. Substitution of Equation (2) in Equation (1) yields

$$\bar{U} = ct^{-m} - St^{-1} \tag{3}$$

The time $t*$ (critical time) for which $\bar{U} = \bar{U}_{max}$ is obtained by taking

$$\frac{d\bar{U}}{dt} = 0 \tag{4}$$

If S is assumed to be constant (storage capacity available), this gives

$$t* = \left(\frac{S}{mc}\right)^{\frac{1}{1-m}} \tag{5}$$

Substitution of Equation (5) in Equation (3) yields

$$\bar{U}_{max} = c(1-m)\left(\frac{S}{mc}\right)^{\frac{-m}{1-m}} \tag{6}$$

$\bar{U}_{max}$ is the average groundwater discharge during the rise of the groundwater table that is required to avoid exceeding the available groundwater storage capacity. For simplicity's sake one assumes that the average discharge is half its highest value during the rise of the groundwater table from the depth where discharge starts, to the surface. In the following the stream spacing will be considered as a function of the required discharge U_r at the highest groundwater level. Thus

$$U_r = 2\bar{U}_{max} \tag{7}$$

The storage capacity $S(d)$ is defined as the volume of water that is stored per unit surface area when the groundwater table rises from a given depth d to the surface; it depends on the storage factor μ. The relationship between groundwater depth d and storage factor μ obtained from various observations by De Vries (1974) was used to formulate the following approximation:

$$u = a(d - z) \tag{8}$$

where a is a constant and z forms the vertical coordinate that is taken positive in a downward direction, so that μ decreases from the surface to the water table.

The storage capacity $S(d)$ between the surface and the groundwater table is

$$S(d) = \int_o^d \mu dz \tag{9}$$

or

$$S(d) = 0.5\, ad^2 \tag{10}$$

The factor a is close to 0.1 m^{-1} and since the maximum value of μ is approximately 0.25, Equation (10) is limited to $d < 2.5$ m. The storage capacity $S(d)$ for 0 m $< d <$ 2.5 m can thus be approximated by

$$S(d) = 0.05\, d^2 \quad (S \text{ in metres}) \tag{11}$$

Assuming a parabolic groundwater table, then the average groundwater head is 2/3 of the head midway between the drainage channels.

The total increase in storage during a rise of a parabolic groundwater table is approximately $0.05(0.66\, d)^2 = 0.022\, d^2$ if d is the groundwater depth midway between the drainage channels (Fig. 2). The total increase in storage during a rise of the groundwater table from the discharge base to the surface is thus:

$$S_n = 0.022\, (d_n^2 - d_{n-1}^2) \tag{12}$$

where d_n represents the incision depth of the streams of nth-order.

A combination of Equations (6), (7) and (12) yields the required discharge U_r as a function of the initial groundwater depth d at the beginning of the rainy period:

$$U_{r,n} = 2c(1-m)\left(\frac{0.022\,(d_n^2 - d_{n-1}^2)}{mc} \right)^{\frac{-m}{1-m}} \tag{13}$$

For Equation (2) the following parameters were applied by De Vries (1974) for the Netherlands conditions:

Frequency	c (mm day^{-1})	m
1%	20	0.38
5%	10	0.25
10%	6	0.175
20%	3	0.05

These figures substituted in Equation (2) give the average amount of rainfall per time unit that is reached or exceeded during a rainfall event, occurring with indicated percentage of all days within a winter period.

2.4 Drainage formulae

Groundwater flow to the stream systems can be schematized as a flow towards a parallel set of drains (Fig. 2), and can be described by the formula given by Ernst (1956):

$$\frac{h_m}{U} = \frac{L^2}{8T} + L\Omega \quad [T] \tag{14}$$

where U is average flux through the phreatic surface per unit area [LT^{-1}], T is transmissivity of the aquifer [L^{2}T^{-1}], h_m is difference in hydraulic head between the divide and the discharge base [L], L is stream spacing [L], Ω is resistance for the radial flow component near the partly penetrating drain [TL^{-1}], caused by the upward bending and contraction of the flow lines.

The radial flow resistance depends on geometry and incision depth of the drainage channel and the ratio between the hydraulic soil properties for the fine Coversand layer and the coarse underlying aquifer material. The ratio of the hydraulic conductivities is about 1:5, whereas the ratio of the thickness for these layers is 1:10. With these figures and modified formulae on radial flow given by Ernst (1956), the following relation was derived by De Vries (1974):

$$\Omega = \frac{1}{\pi K'} \ln \frac{5b'}{B} \quad [TL^{-1}] \tag{15}$$

where K' [LT^{-1}] and b' [L] are the hydraulic conductivity and thickness of the Coversand layer, and B is the wetted perimeter of the stream [L].

2.5 Streamflow formulae

Streamflow can be described by the well known Gauckler-Manning formula:

$$Q = k_m A R^{0.67} s^{0.5} \quad [L^3 T^{-1}] \tag{16}$$

where A is wetted cross section [L^2], k_m is roughness coefficient [$L^{0.33}T^{-1}$], R is hydraulic radius (A/B) [L], s is hydraulic gradient (slope of stream bed for uniform flow); Q = discharge [$L^3 T^{-1}$], B is wetted perimeter [L].

If we assume for simplicity's sake half-circular channels with radius r, then: $A = 0.5$ πr^2 and $B = \pi r$, thus $R = 0.5r$.

Substitution of these relations in Equations (14)-(16) for the maximum groundwater discharge gives:

$$U_{max} = \frac{h_m}{\dfrac{L^2}{8T} + \dfrac{L}{\pi K'} \ln \dfrac{5b'}{\pi r}} \quad [LT^{-1}] \tag{17}$$

where for the 1st-order system $h_{m,1} \approx d_1$, and for higher order systems $h_{m,n} \approx d_n - d_{n-1}$. U_{max} equals U_r from Equation (13), and

$$Q = k_m r^{2.67} s^{0.5} \quad [L^3 T^{-1}] \tag{18}$$

The discharge *(Q)* at the junction of the stream of next higher order equals (assuming parallel streams of the same order and orthogonal confluences):

$$Q = ULl \tag{19}$$

where l is stream length. If α is introduced as the ratio l/L and one applies in further considerations the discharge midway between two junctions *(Q)* as the flow rate controlling the average stream dimensions, then one has

$$Q = 0.5 \, \alpha \, U L^2 \tag{20}$$

Substitution of Equation (18) gives

$$U_{max} = \frac{k_m r^{2.67} s^{0.5}}{0.5 \, \alpha \, L^2} \quad [LT^{-1}] \tag{21}$$

For the 1st-order system $U_{max,1}$ in Equation (21) is equal to $U_{r,1}$ in Equation (13).

The higher order streams must also be able to discharge as stream flow the discharge from the 1st-order system.

3   RESULTS

For different precipitation characteristics (as distinguished by their frequency of occurrence), the required discharge and the appropriate stream spacing and associated channel size were calculated from Equations (13), (17) and (21) for given incision depth, stream bed slope, channel roughness and subsurface permeability. The results are depicted in Figure 3 for a stream network consisting of streams of two orders for average conditions in the Netherlands Pleistocene area (Table 1).

It was tentatively assumed that the first and second order stream systems have respectively been adapted to a rainfall frequency of 5% and 20%. The threshold value of 5% for the natural system was tentatively chosen because in cases where the drainage conditions have been improved artificially, a design criterium of about 1% is applied in the Netherlands.

The incision depth is predominantly connected with the primary topography, that means with the geomorphological evolution that belongs to a larger time and areal scale than the considered stream system. Therefore, the incision depth in the present study is considered as a semi-independent variable.

Figure 3 shows that the first order stream spacing increases from 150 m to 800 m due to an increase in incision depth from 0.5 m to 1.5 m. The channel radius then increases from 0.3 m to 1 m and the discharge rate reduces from 9 mm day^{-1} to 5 mm day^{-1}. The second order system reaches a maximum spacing of 1000 m and a channel

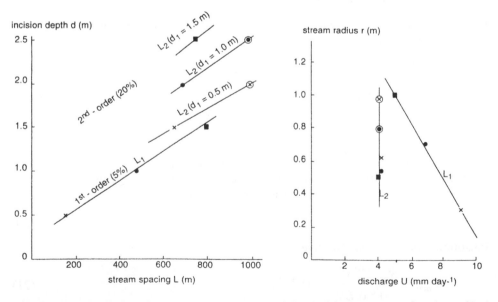

Figure 3. First and second-order stream spacing, channel size and drainage rate as functions of incision depth. Calculated from Equations (13), (17) and (21) with data from Table 1.

Table 1. Average climatic, geomorphological and geological parameters for the Netherlands for first and second order drainage systems according to De Vries (1974) and Ernst (1978).

	1st-order	2nd-order
Transmissivity T (m^2 day^{-1})	1000	1000
Permeability upper layer K' (m day^{-1})	3	3
Thickness upper layer b' (m)	5	5
Rainfall frequency (% exceedance)	5	20
Stream bed slope s	1:500	1:1500
Channel-roughness factor k_m (m$^{0.33}$ sec^{-1})	20	30
Stream spacing/stream length ratio α	0.25	0.1

radius of 1 m for an incision depth of 2.5 m, in the case of a first-order stream incision depth of 0.5 m.

An increase in the first-order incision depth to 1.5 m results in a reduction of the maximum second-order stream spacing to 700 m, that means that the first and second-order systems coincide, or rather that the second-order system will not develop. The second-order discharge does not vary much with incision depth and is close to 4 mm day^{-1}.

4 DISCUSSION AND CONCLUSIONS

1. Stream spacing and channel size increase almost linearly with incision depth, until a depth of about 2.5 m. Above this depth the total annual precipitation surplus can be stored and the required discharge becomes independent of the groundwater depth and reduces to the average annual precipitation surplus of about 1 mm day^{-1}.

2. The bifurcation ratio L_n / L_{n-1} reduces with a decrease in difference between first and second-order incision depths, that means with a decrease in relief. First and second-order streams coincide if the first-order incision depth becomes close to 1.5 m. That means that the storage capacity is large enough to reduce the required discharge during a rainfall event which occurs with a frequency of 5% to the required discharge for a 20% event.

3. The required discharge capacity of the second-order system is hardly dependent on the incision depth and is close to 4 mm day^{-1}. This is due to the low gradient of the rainfall intensity-duration curve: the exponent m in Equation (2) is 0.05. This means that the seasonal groundwater level fluctuation (change in storage) is more or less independent of the incision depth and the stream spacing. Consequently, a larger spacing (and thus a greater hydraulic head h_m) means that the ratio between seasonal fluctuation and average head reduces, so that the discharge becomes more equally distributed over the wet and dry seasons.

4. The calculated stream spacings agree in order of magnitude with the observed situation in the Netherlands Pleistocene area, and show the same tendency of increasing spacing with increasing average groundwater depth (Fig. 1). The calculated discharge capacities also agree well with the values observed by De Vries (1974) in the Netherlands in adequately dewatered areas where the groundwater discharge capacity is exceeded not more often than in 1% of events in a winter period. These values decrease

from about 12 mm day^{-1} to 3 mm day^{-1} if the groundwater depth increases from 0.4 m to 1.4 m.

5. There is some discrepancy between the observed actual stream spacing in the Netherlands in an absolute sense and the theoretical calculations. This is due to the artificial character of part of the drainage system. The implications of this artificially improved dewatering has been discussed in De Vries (1994).

In conclusion: The proposed conceptual model illustrates in a quantitative way the functional relationship between the geological and climatic conditions on one side and the structure and behavior of the drainage network on the other. This model can be applied to appraise the influence of environmental change on the stream network and the drainage conditions.

'After this paper was submitted to WDU a more extended version has been published (De Vries, 1995)'.

REFERENCES

De Vries, J.J. 1974. Groundwater flow systems and stream nets in the Netherlands. Ph.D. thesis, Vrije Universiteit, Amsterdam; Editions Rodopi, Amsterdam.

De Vries, J.J. 1994. Dynamics of the interface between streams and groundwater systems in lowland areas, with reference to stream net evolution. *J. Hydrology* 155: 39-56.

De Vries, J.J. 1995. Seasonal expansion and contraction of stream networks in shallow groundwater systems. *J. Hydrology* 170: 15-26.

Ernst, L.F. 1956. Calculations of the steady flow of groundwater in vertical sections. *Neth. J. Agric. Sci.* 4: 126-131.

Ernst, L.F. 1978. Drainage of undulating sandy soils with high groundwater tables. *J. Hydrology* 39: 1-50.

CHAPTER 6

Field validation of a water and solute transport model for the unsaturated zone

W.J. BOND & C.J. SMITH
CSIRO Division of Soils, Canberra, Australia

P.J. ROSS
CSIRO Division of Soils, PMB, Aitkenvale, Australia

ABSTRACT: Models that are to be used for describing natural processes require validation prior to their widespread use to ensure that they describe those processes adequately. Validation should be carried out by comparing field measurements with model predictions based on independently measured input data, without any optimisation of the observations. This has rarely been achieved for the movement of water and solute through the unsaturated zone. This paper describes the validation of a model based on numerical solutions of Richards' equation and the advection-dispersion equation. Model predictions were compared with average bromide distributions obtained at three sampling times during a six month period of natural rainfall after surface application. The model was provided with independently measured soil hydraulic properties, rainfall and evaporation data. Dispersion was described by the molecular diffusion coefficient with tortuosity set to one and dispersivity set to zero. Good agreement between model predictions and measured bromide distributions was observed. The model predicted adequately the peak location, spread and asymmetry of the bromide pulse at each of the sampling times. No additional mechanism such as preferential flow was necessary to predict the solute pulse asymmetry. This study provides some confidence that traditional soil physics theory can be applied successfully to describe water and solute transport in at least some field soils.

1 INTRODUCTION

Models of soil water flow and solute transport are required to assist the understanding of processes in the unsaturated zone, to interpolate and extrapolate experimental findings, and to make predictions and evaluate scenarios. Before this is possible there needs to be some level of confidence that the model adequately represents field-scale processes, that is to say the model needs to be validated. Validation has been interpreted in a number of ways and is often confused with calibration. In this paper we take validation to mean the comparison of field measurements with model predictions based on independently measured input parameters. This contrasts with calibration, which usually refers to the fitting of model predictions to field measurements by adjusting the input parameters.

By the above criteria, there have been very few reports in the literature of the validation of solute transport models under realistic transient field-scale conditions. Butters & Jury (1989) stated that no deterministic one-dimensional dispersion model has been able to describe field scale transport accurately. This had earlier been the justification for the development of non-mechanistic models, such as the transfer function model of Jury (1982). Tests of such models, which do not use input parameters measured independent of the experiment against which the tests are carried out, fall into the calibration category (e.g. Jury et al., 1982; Jury & Sposito, 1985; Butters & Jury, 1989).

Comfort et al. (1993) tested the one-dimensional mechanistic LEACHM model of Wagenet & Hutson (1989) against field measurements of bromide movement. They showed that LEACHM adequately predicted bromide transport when provided with independently assessed input parameters. Hills et al. (1991) found that an independently parameterised mechanistic, two-dimensional unsteady solute transport model provided only qualitative agreement with field measurements of bromide and tritium movement at the Las Cruces Trench Site. Other field tests of mechanistic water and solute transport models have been calibrations, where the model has been fitted to the data (e.g. Jury & Sposito, 1985; Jaynes et al., 1988; Butters & Jury, 1989; Jaynes, 1991; Roth et al., 1991). In such tests it has often been found that mechanistic models were unable to describe solute movement satisfactorily. This may be because of the use of models that were inappropriate to the boundary conditions of the field experiment to which they were applied. An example of this is the application of a steady state solute transport model to experiments where flow was (sometimes dramatically) unsteady (e.g. Butters & Jury, 1989; Roth et al., 1991).

This paper reports the successful field validation of a general mechanistic model for unsteady unsaturated water and conservative solute movement through soils, based on the numerical solution of Richards' equation and the advection-dispersion equation. Validation was carried out against bromide leaching measured at intervals over a six month period of winter and spring rainfall. All input parameters, including hydraulic properties and dispersion coefficient estimates, were determined independently from the field measurements used for validation.

2   METHODS

This work forms part of larger study of the irrigation of tree plantations with sewage effluent (the Wagga Effluent Plantation Project), which includes estimation of deep drainage and solute leaching by measurement and modelling. The validation experiment was carried out on one part of the field site being used for this project near Wagga Wagga in New South Wales, Australia, namely the 1.5 ha block planted with *P. radiata*. This block is divided into eight plots which represent two replicates of each of four different irrigation treatments (Fig. 1). For the purpose of the present study, the different treatments can be ignored. The bromide transport experiment was carried out at the end of the first irrigation season after the commencement of winter rainfall. At that time the soil profile water storage, as a percentage of saturation, was

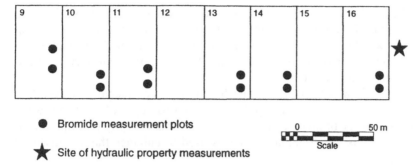

Figure 1. Location of bromide tracer sub-plots and the soil hydraulic property measurements in the *P. radiata* block at the Wagga Effluent Plantation Project site.

uniform across the block. The soil at the site is classified as an alfisol (US Taxonomy) or a mixture of Red Earths and Red Podzolics (Great Soil Group). There is a texture change from sandy loam to sandy clay or light medium clay occurring at between 0.5 and 0.8 m and there is often a repeated cycle of coarse and fine texture below that.

2.1 *Bromide transport measurements*

Bromide, in the form of a 50 mol m^{-3} solution of KBr, was applied to 12 circular sub-plots, each 1 m^2 in area, distributed across the *P. radiata* block as shown in Figure 1. The KBr solution was injected into the soil at each point on a 50 mm grid at a depth of 20 mm using an automatically re-filling animal vaccination gun connected to a reservoir containing the KBr solution. Two mL of solution was injected at each point. Injection was carried out four times at weekly intervals starting on 26 May 1992. This was not necessary for the validation experiment but was carried out because an experiment involving ^{15}N was being carried out simultaneously and the latter required an application distributed over three weeks. The resulting Br application rate was 0.16 mol m^{-2}.

At the time of the experiment the trees, which were planted on a 2 × 3 m spacing, were less than a year old and had an average height of 1.3 m. Their root systems were still restricted and were not expected to influence the fate of the tracers in the subplots equidistant from trees. Grass covered the surface between the trees, but the plots were kept bare for the duration of the experiment by regular spraying with the herbicide glyphosate.

The sub-plots were destructively sampled on three occasions to determine the distribution of bromide with depth. The first sampling time was one week after the final application, 28 days after the initial application, on 23 June. The second and third samplings were on 21 July and 25 November, 8 weeks and 26 weeks after initial application, respectively. At each time two continuous cores (42 mm diameter) to 0.9 m were collected at a distance of 0.15 m and 0.43 m from the centre of each sub-plot. Two additional cores were taken 0.15 m and 0.43 m from the outside edge of the sub-plots to check for lateral movement of the tracers outside the area of application. All holes created by coring were repacked with subsoil material. Each core was sectioned into the depth intervals 0 to 0.05, 0.05 to 0.1, 0.1 to 0.2, 0.2 to 0.3, 0.3 to 0.4, 0.4 to 0.5, 0.5 to 0.7, and 0.7 to 0.9 m. All the soil from each depth interval was retained, and was weighed, dried, ground, mixed and sub-sampled for analysis of bromide and determination of the water content. This permitted an accurate mass balance to be calculated for each core.

A 4 g sub-sample of soil was taken from each section and extracted with 20 mL of distilled water. The extract was centrifuged and filtered through a 0.45 μm filter to produce a clear sample for bromide analysis by ion chromatography (Millipore Corporation, 1989). A uv-vis spectrophotometer was used to enhance the sensitivity of bromide detection.

The concentration of the soil extract was used to calculate the Br concentration of the soil as mol kg^{-1}. This was then used together with the dry weight of soil to calculate the mass of Br in each depth interval and the total mass of Br in each core, in order to determine the mass balance of bromide. It was also used together with the water content to calculate the concentration of Br in the soil solution at the time of sampling in each section of each core.

2.2 Model description

The model being validated is based on the SWIM model for water movement through unsaturated soils (Ross, 1990a), which uses efficient methods (Ross, 1990b) for solution of Richards' equation, Equation (1):

$$\frac{\partial \theta(\psi)}{\partial t} = \frac{\partial}{\partial z}\left\{K(\theta)\left[\frac{\partial \psi}{\partial z} - 1\right]\right\} \tag{1}$$

where θ is water content, ψ is the soil water potential, K is the soil hydraulic conductivity, t is time and z is depth. A finite difference solution of the advection-dispersion equation (Eq. 2) has recently been incorporated into this model:

$$\frac{\partial(\theta C)}{\partial t} = \frac{\partial}{\partial z}\left[\theta D_s \frac{\partial C}{\partial z}\right] - \frac{\partial(vC)}{\partial z} \tag{2}$$

where C is solute concentration in the soil solution, v is the soil water flux [= $K(\partial\psi/\partial z - 1)$], and D_s is the solute dispersion coefficient. Comparison against analytical solutions for both steady and unsteady flow conditions has shown that the model's correction for numerical dispersion eliminates numerical dispersion for most flow velocities in field soils. The SWIM model has been successfully used to describe water movement by Bond et al. (1994).

2.3 Measurement of soil hydraulic properties

The unsteady drainage flux method, similar to that described by Green et al. (1986), was used to determine the hydraulic properties required by the model. The measurements were made in duplicate at a single site adjacent to the *P. radiata* block (Fig. 1). Soil water retention curves [$\psi(\theta)$] and hydraulic conductivity – water content relationships [$K(\theta)$]were obtained at 0.1 m intervals from 0.1 m to 1.5 m. Examples of the $\psi(\theta)$ and $K(\theta)$ data are shown in Figure 2 for four representative depths.

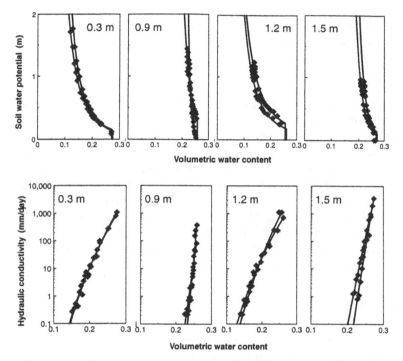

Figure 2. Water retention curves, $\psi(\theta)$, and hydraulic conductivity – water content relationships, $K(\theta)$, for four representative depths. Points are replicate measurements, lines are fitted using Equations (3) and (4).

The soil water retention data were described very well by the Campbell (1974) form of the Brooks & Corey (1964) equation,

$$(\theta/\theta_s) = (\psi/\psi_e)^{-1/b} \qquad \psi < \psi_e$$
$$\theta = \theta_s \qquad\qquad\qquad \psi > \psi_e \tag{3}$$

where θ is the volumetric water content, θ_s is the water content at saturation, ψ_e is the air entry water potential and b is an empirical constant. The discontinuity at $\psi = \psi_e$ was avoided by using the smoothing function of Hutson & Cass (1987). Values of θ_s were obtained from the water content measurements prior to drainage. Similarly, the hydraulic conductivity – water content data were described well by a generalised form of the equation of Brooks & Corey (1964) and Campbell (1974), namely:

$$(K/K_s) = (\theta/\theta_s)^{bm} \tag{4}$$

where K_s is the saturated hydraulic conductivity and m is an empirical constant related to the pore interaction factor p by $m = 2 + (2+p)/b$ (Campbell, 1974). Equation (4) was fitted to the data without normalisation and the parameters obtained from this fit were used to calculate m and K_s, the latter from the known value of θ_s. This extrapolated value of K_s, based on matrix properties, does not account for any contribution of macropores close to saturation. This is thought to be unimportant, however, because the hydraulic properties of the soil are such that saturation is rarely approached for the rainfall rates it is usually subjected to.

2.4 *Other model inputs*

Daily rainfall and pan evaporation figures, obtained from a nearby meteorological station, were used. For the bare soil plots used in these experiments, potential evaporation was assumed to be 70% of evaporation from a class A pan (cf. Penman, 1948). Cumulative values for the periods between samplings are presented in Table 1. Each day's rainfall or evaporation was assumed to occur with a constant average rate over a 12 hour period. For rainfall, this had the effect of preventing runoff that might have been caused by high instantaneous rainfall rates, which for this soil with a high surface infiltration rate was not of concern. The development of a surface crust, which is one of the available features of the SWIM model, was de-activated.

The model can accept the general form of the hydrodynamic dispersion coefficient used by many authors (e.g. Bond, 1986), namely:

$$D_s = \gamma D_0 + \alpha \left(\frac{v}{\theta} \right)^n \tag{5}$$

where D_0 is the molecular diffusion coefficient in free water, γ is the tortuosity, α is the dispersivity, and n is another empirical constant. Because the average net infiltration rate during the experiment was only 2.2 mm day^{-1} (Table 1) the dispersivity was set to zero. The tortuosity was set to unity because laboratory experiments have shown that this provides a good description of dispersion during transport (Bond, 1987). The diffusion coefficient of KBr in water was set to 2×10^{-9} m s^{-1}, after Robinson & Stokes (1959).

2.5 *Elimination of anion exclusion*

Anion exclusion is commonly observed to affect anion transport in clay soils (Bond et al., 1984). Laboratory experiments were carried out using both topsoil and subsoil samples from the experiment site to determine whether Br was affected by exclusion in this soil. The experiments were similar to those described by Bond et al. (1982). A solution of NaBr (5 mol m^{-3}) in a background of 5 mol m^{-3} NaCl and 5 mol m^{-3} CaCl$_2$ was supplied at a low, constant rate to an initially moist but unsaturated column of soil. The concentrations and the cation composition of the inflowing solution

Table 1. Cumulative rainfall and evaporation from a class A pan for different time intervals during the bromide tracer experiment.

	Rainfall (mm)	Pan evaporation (mm)	70% Pan evaporation (mm)
26 May-23 June	53.8	22.6	15.8
23 June-21 July	42.0	25.4	17.8
21 July-25 Nov	355.0	356.0	249.2
Total	450.8	404.0	282.8

were chosen to match those observed during the field experiment. When the wetting front had penetrated a distance of 0.25 m, the column was divided into 8 mm sections. The water content was determined in each section and each was analysed for Br^-, Cl^- and NO_3^- by ion chromatography following extraction with 20 mL of 10 mol m^{-3} $MgSO_4$ (to encourage flocculation). In both the topsoil and subsoil samples, the distinct 'front' between the concentration of each ion in the inflowing solution and that in the initial soil solution was found at the same distance into the soil column for all three anions measured. By comparing this location with the piston front of the water (Bond et al., 1982) it was concluded that there was no measurable enhancement of anion transport relative to the water. Anion exclusion could therefore be omitted from the model.

3 RESULTS AND DISCUSSION

3.1 *Preliminary model test*

As a preliminary test of the ability of the model to describe water movement in a vertically heterogeneous profile, it was used to predict the drainage behavior observed during the measurement of the hydraulic properties by the unsteady drainage flux method. Although this does not constitute validation it is one step in testing the performance of the model. The initial water contents were set to saturation, evaporation and rainfall were set to zero, and the model was used to simulate the evolution of the water content profile. In Figure 3 the measured and simulated water content profiles after 28 hours and 30 days are compared. Agreement is good, showing that if an adequate set of hydraulic properties is provided, the model can describe unsaturated water movement in this complex soil profile.

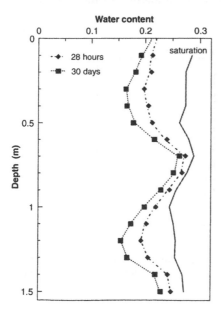

Figure 3. Measured (points) and predicted (lines) water content profiles at two times after commencement of drainage from saturation.

3.2   *Measured bromide distributions*

The bromide distributions obtained from each core were examined individually before being averaged. Each core contained a single bromide peak, and although the
spread and peak position varied slightly between cores there was no evidence of
preferential movement of some fraction of the bromide. The average distributions at
each sampling time are presented in Figure 4 as the plus and minus 95% confidence
limits about the mean concentration. The 95% confidence interval is quite small indicating that the amount of variability between the 24 cores collected from the 12
sub-plots is not large. This makes the data set suitable for a model validation procedure.

It is obvious from Figure 4 that the bromide pulse had moved beyond the sampling range by the third sampling. This is confirmed by the average mass of bromide
recovered from the cores which was 84% at the first sampling, 89% at the second
sampling, and only 18% for the third sampling time. Lack of complete mass balance
at the first two samplings is attributed to the inevitable losses associated with lateral
movement away from the plot area, uptake by plants that re-grew between herbicide
treatments, and loss of soil during sampling. Examination of the cores taken outside
the bromide application area showed that although no bromide was detectable for the
first two samplings, by the third sampling bromide had moved laterally, further contributing to the poor recovery from the cores inside the application area at that time.

3.3   *Model predictions*

The bromide distributions predicted by the model are also shown in Figure 4. These
predictions were not optimised to the data in any way. The agreement of the predictions with the measurements is very good. The prediction of the bromide peak position confirms that the SWIM model accurately predicted the average soil water flux
over the time intervals and depths tested. The accurate prediction of the shape of the

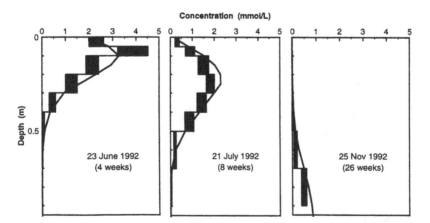

Figure 4. Average bromide distributions at the three sampling times. Shaded areas represent the
95% confidence intervals about the mean for each depth interval. The solid lines are the model predictions.

bromide distributions confirms the adequacy of the model's solution of the advection-dispersion equation for describing the spreading of a pulse of bromide during transient soil water flow in this soil.

Good independent predictions of field-scale water and solute transport using Richards' equation and the advection-dispersion equation have not often been achieved in the past. Possible explanations for why other studies have failed to validate such models include the difficulty in establishing experimental boundary conditions that match the model, and the inappropriate use of steady-state models to describe unsteady transport. It is recognised that the observations for this soil cannot necessarily be generalised to other soils. However, it is stressed that the soil used here was not selected specifically for the purpose of validation because of any special properties, but is a typical soil in the area where it is found. One important feature of water flow in this soil which may improve the likelihood of agreement with model predictions is that the soil rarely becomes saturated. This means that preferential flow, which is much more likely at or near to saturation, will not have contributed to determining the fate of solutes to the extent that it may have in some other studies.

3.4 *Pulse asymmetry*

It is particularly interesting to note that the asymmetry of the bromide distribution, which is obvious at the first two sampling times, is adequately predicted by the model. It is often suggested that asymmetric distributions are the result of preferential flow of water and solutes through macropores (e.g. Wild & Babiker, 1976). We have shown here, however, that such distributions can be predicted from the advection-dispersion equation solved for unsteady flow conditions. Part of the asymmetry can be ascribed purely to the effect of the non-uniform water content (cf. Wilson & Gelhar, 1981). Examination of the evolving bromide distributions predicted by the model showed that the bulk of the effect is a result of the alternating wetting and drying occurring at the soil surface. During a drying phase, solute close to the soil surface was drawn upwards and accumulated at the surface. The trailing edge of the solute pulse was thereby compressed and sharpened. Meanwhile the leading (deeper) edge of the pulse continued to spread by diffusion. When the next wetting event occurred, the trailing edge of the pulse moved into the soil and recommenced spreading by diffusion. A long time of downward movement was required, however, before the trailing edge spread as much as the leading edge and the pulse approached a symmetrical shape. A process similar to this was described by Bond (1987) in laboratory columns, where different times available for diffusion at the leading and trailing edges of a pulse were shown to cause peak asymmetry.

4 CONCLUSIONS

It has been shown that a model, based on a Richards' equation description of water flow and the advection-dispersion equation for solute transport, and provided with independently measured input parameters, satisfactorily predicted measured distributions of bromide obtained from carefully controlled field experiments. There was no fitting of the model to the data or any other form of optimisation. The asymmetric bromide distributions were quite adequately described by the model without any need to invoke preferential flow. The spreading of the pulse could be described adequately with dispersivity set to zero, thus ignoring all spreading mechanisms except molecular diffusion. This is probably because of the low average pore water velocity.

This study provides some confidence that traditional soil physics theory can be applied successfully to describe water and solute transport in at least some field soils. Work is in progress to obtain data sets at deeper depths and under other conditions in order to ensure that good prediction is not limited to the current example.

ACKNOWLEDGMENTS

We are pleased to acknowledge the experimental assistance of Evelyn Johnson, Gordon McLachlan and Seija Tuomi, and helpful discussions with Kirsten Verburg. This work was partially funded by the Land and Water Resources Research and Development Corporation, the Murray Darling Basin Commission, and the NSW Public Works Department.

REFERENCES

Bond, W.J. 1987. Solute transport during unsteady, unsaturated soil water flow: the pulse input. *Aust. J. Soil Res,* 25: 223-241.

Bond, W.J. 1986. Velocity-dependent hydrodynamic dispersion during unsteady, unsaturated soil water flow: experiments. *Water Resour. Res.* 22: 1881-1889.

Bond, W.J., Gardiner, B.N. & Smiles, D.E. 1982. Constant flux absorption of a tritiated calcium chloride solution by a clay soil with anion exclusion. *Soil Sci. Soc. Am. J.* 46: 1133-1137.

Bond, W.J., Gardiner, B.N. & Smiles, D.E. 1984. Movement of $CaCl_2$ solutions in an unsaturated clay soil: the effect of solution concentration. *Aust. J. Soil Res.* 22: 43-58.

Bond, W.J., Willett, I.R. & Verburg, K. 1994. Determining the fate of contaminants following land disposal of mine waste water. *Proc. XV Congress, ISSS,* vol. 3b, 425-426, Acapulco, Mexico, July 1994.

Brooks, R.H. & Corey, A.T. 1964. Hydraulic properties of porous media. *Hydrology paper* No. 3, Colorado State Univ., Fort Collins.

Butters, G.L. & Jury, W.A. 1989. Field scale transport of bromide in an unsaturated soil, 2. Dispersion modeling. *Water Resour. Res.* 25: 1583-1589.

Campbell, G.S. 1974. A simple method for determining unsaturated conductivity from moisture retention data. *Soil Sci.* 117: 311-314.

Comfort, S.D., Inskeep, W.P. & Lockerman, R.H. 1993. Observed and simulated transport of a conservative tracer under line-source irrigation. *J. Environ. Qual.* 22: 554-561.

Green, R.E., Ahuja, L.R. & Chong, S.K. 1986. Hydraulic conductivity, diffusivity, and sorptivity of unsaturated soils: Field methods. In: A. Klute (ed.), *Methods of Soil Analysis. Part 1. Physical and mineralogical methods,* (2nd ed): 771-798.

Hills, R.G., Wierenga, P.J., Hudson, D.B. & Kirkland, M.R. 1991. The second Las Cruces Trench experiment: Experimental results and two-dimensional flow predictions. *Water Resour. Res.* 27: 2707-2718.

Hutson, J.L. & Cass, A. 1987. A retentivity function for use in soil-water simulation models. *J. Soil Sci.* 38: 105-113.

Jaynes, D.B. 1991. Field study of bromacil transport under continuous-flood irrigation. *Soil Sci. Soc. Am. J.* 55: 658-664.

Jaynes, D.B., Rice, R.C. & Bowman, R.S. 1988. Independent calibration of a mechanistic-stochastic model for field-scale solute transport under flood irrigation. *Soil Sci. Soc. Am. J.* 52: 1541-1546.

Jury, W.A. 1982. Simulation of solute transport using a transfer function model. *Water Resour. Res.* 18: 363-368.

Jury, W.A. & Sposito, G. 1985. Field calibration and validation of solute transport models for the unsaturated zone. *Soil Sci. Soc. Am. J.* 49: 1331-1341.

Jury, W.A., Stolzy, L.H. & Shouse, P. 1982. A field test of the transfer function model for predicting solute transport. *Water Resour. Res.* 18: 369-375.

Millipore Corporation. 1989. *Waters Ion Chromatography Cookbook.* Millipore Corporation, Waters Chromatography Division, Ion Chromatography Group, Massachusetts.

Penman, H.L. 1948. Natural evaporation from open water, bare soil and grass. *Proc. Royal Soc. A,* 193: 120-145.

Robinson, R.A. & Stokes, R.H. 1959. *Electrolyte solutions,* 2nd ed. London: Butterworths.

Ross, P.J. 1990a. Efficient numerical methods for infiltration using Richards' equation. *Water Resour. Res.* 26: 279-290.

Ross, P.J. 1990b. SWIM – A simulation model for soil water infiltration and movement, Reference Manual, CSIRO Division of Soils, Australia.

Roth, K., Jury, W.A., Flühler, H. & Attinger, W. 1991. Transport of chloride through an unsaturated field soil. *Water Resour. Res.* 27: 2533-2541

Wagenet, R.J. & Hutson, J.L. 1989. LEACHM: Leaching estimation and chemistry model. Continuum Vol. 2, Water Resour. Inst., Cornell Univ., Ithaca, NY.

Wild, A. & Babiker, I.A. 1976. The asymmetric leaching pattern of nitrate and chloride in a loamy sand under field conditions. *J. Soil Sci.* 27: 460-466.

Wilson, J.L. & Gelhar, L.W. 1981. Analysis of longitudinal dispersion in unsaturated flow. 1. The analytical method. *Water Resour. Res.* 17: 122 – 130.

CHAPTER 7

Characterization of a regional groundwater discharge area by combined analysis of hydrochemistry, remote sensing and groundwater modelling

O. BATELAAN & F. DE SMEDT
Free University Brussels, Laboratory of Hydrology, Belgium

P. DE BECKER & W. HUYBRECHTS
Institute of Nature Conservation, Hasselt, Belgium

ABSTRACT: A multi-layer regional groundwater flow model for simulation of quantitative groundwater discharge has been developed and applied to the wetland nature reserve Walenbos, (Belgium). The model calculates groundwater seepage, infiltration zones, fluxes and travel times in an interpretative sense. The different vegetation types and chemical characteristics of the seepage water in the nature reserve compared very well with the different local and regional groundwater systems simulated by the model. In addition, Landsat TM remote sensing analyses were used to identify and compare the distribution of recharge and discharge areas. All analyses were integrated in a GIS resulting in an efficient multi-disciplinary method for a more fundamental understanding of groundwater wetland flow systems.

1 INTRODUCTION AND GEOGRAPHICAL SETTING

Apart from the basic conditional geological description, most studies analyze regional groundwater systems by studying topography, land and soil features, hydrochemistry, environmental isotopes, piezometric patterns or groundwater numerical model results. Those aspects were listed by Freeze & Cherry (1979) as different ways of mapping discharge and recharge areas. Simultaneous studies of those aspects is often limited due to the tendency for mono-thematic specialisations and lack of data or a platform for integration. With Geographical Information Systems (GIS) an appropriate environment for analysis and comparison of thematic data is now available.

In this paper results of a multi-thematic approach to the analyses of groundwater discharge areas is presented. It is shown and advocated that a multi-thematic GIS-integrated approach results in a better, more holistic understanding of groundwater flow systems.

The approach is applied to the surroundings of the forest nature reserve, Walenbos, located about 30 km east of Brussels, Belgium. The study area has a size of 12 by 16 km (Fig. 1). The topography is characterized by a series of SW-NE undulating Pliocene hills, between 50 and 100 m high, consisting of glauconite rich sand,

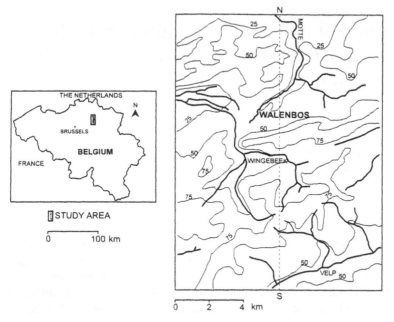

Figure 1. Geographical setting.

often with a hard pan at the top. Walenbos lies in the valley of the River 'Brede Motte'. A slightly elevated ridge divides Walenbos into an eastern and a western sub-basin.

2   DESCRIPTION OF A MULTI-LAYER GROUNDWATER DISCHARGE MODEL

A steady state quasi three-dimensional model was used to simulate the groundwater flow. The hydrogeological conditions consist of a phreatic aquifer, separated by a semi-conducting layer from an underlying semi-confined aquifer. In this layered system the groundwater flow can be described by a horizontal flow equation in the upper phreatic aquifer, with leakage losses through the underlying semi-pervious layer to the semi-confined aquifer, where the flow is also predominantly in the horizontal direction, i.e.

1. Horizontal flow in the phreatic aquifer is described by:

$$\frac{\partial}{\partial x}\left[K(h-z)\frac{\partial h}{\partial x}\right] + \frac{\partial}{\partial y}\left[K(h-z)\frac{\partial h}{\partial y}\right] + c(g-h) + N - P = Q \qquad (1)$$

where K = hydraulic conductivity of the aquifer [LT^{-1}], h = position of water table [L], z = elevation of base of aquifer [L], c = hydraulic conductance of the semi-confining layer [T^{-1}], g = piezometric level of the semi-confined aquifer [L], N = effective precipitation or groundwater recharge [LT^{-1}], P = pumping rate [$L^3T^{-1}L^{-2}$], and Q = groundwater discharge [$L^3T^{-1}L^{-2}$].

2. Horizontal flow in the semi-confined aquifer is described by:

$$\frac{\partial}{\partial x}\left(T\frac{\partial g}{\partial x}\right)+\frac{\partial}{\partial y}\left(T\frac{\partial g}{\partial y}\right)+c(h-g)-P=0 \tag{2}$$

where T = transmissivity [L^2T^{-1}].

Equation (1) needs some more explanation. The identification of groundwater systems involves division of the area into infiltration (recharge) and seepage (discharge) zones. The infiltration zones are areas where there are no rivers or streams and, hence, the seepage is zero ($Q = 0$). Seepage zones are defined as areas where the groundwater emerges to the soil surface such that groundwater discharge occurs. From field considerations these zones were defined as typical wetlands and river valleys, where the groundwater table is near to the soil surface, manifested by the occurrence of springs, rivulets, ditches, etc. In practice, seepage zones are determined in the computer model as zones where the groundwater table reaches within 0.5 m below the soil surface (which is an average watertable depth for wetlands in the vicinity of Walenbos). Note that in this approach it is not necessary to know explicitly the position of streams or wetlands, because the model automatically detects such areas by using the topography (minus 0.5 m) as a limiting criterion.

Additionally, a flow tracking module was developed, based on the Dupuit-Forchheimer assumption, in the sense of neglecting the resistance to vertical flow rather than the flow itself (Strack, 1984). This procedure calculates three-dimensional flow lines starting from each grid cell. Horizontal flow velocity components are calculated from the slope of the watertable or piezometric level. The vertical flow components are interpolated linearly with depth in the top and bottom aquifers and are assumed constant in the aquitard, as illustrated in Figure 2.

The model was integrated in the GIS GRASS at a library level by Batelaan et al. (1993). The model has been applied to the nature conservation area of Walenbos. The hydrogeological situation is schematized by Bronders (1989) as:

1. A phreatic aquifer of 4 to 95 m thickness, in the Formation of Diest, consisting mainly of Pliocene, glauconite-rich sand with a hydraulic conductivity of 11 m day^{-1};

2. A semi-confining 30 m thick layer, in the Rupel and Tongeren Formation, mainly consisting of clay and sandy clay with a hydraulic resistance of 0.007 day^{-1};

3. A semi-confined aquifer, in the Formation of Brussels, consisting of 15 m coarse sands with an average hydraulic conductivity of 7 m day^{-1}.

The situation is shown in Figure 3 by a N-S cross section.

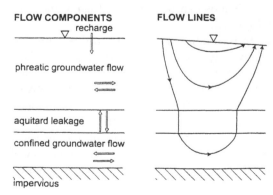

Figure 2. Sketch of velocity components and flow lines.

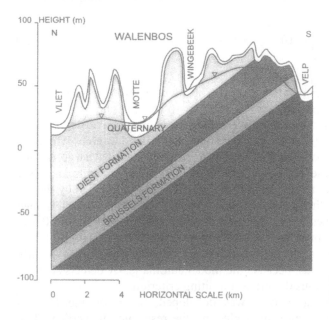

Figure 3. Hydrogeological N-S profile through the study area.

The equations are solved numerically with a finite difference scheme using a five point stencil and a red-black ordered Gauss-Seidel iterative solution procedure (Hackbusch, 1985). The area of 12 km by 16 km was discretized by a grid of computational cells of 50 m by 50 m. The total number of cells amounts to 76,800. It is assumed that the study area has impervious boundaries. This is clearly a gross simplification, but because the calculation procedure implies the automatic identification of infiltration and seepage areas, based on the known topography, a strong conditioning of the groundwater pattern occurs. As it is known from field investigations that there are abundant rivers and rivulets in the area, it follows that in the central part of the study area the simulation results will resemble the actual situation very well. At the boundary, deviations will occur but as can be seen from the following results, these are rather far from the target Walenbos area and its infiltration area and, hence, have no major effect on the results.

Figure 4 shows the resulting groundwater discharge areas. Within the Walenbos two distinct groundwater discharge areas appear. For each of these, the contributing infiltration areas and the flow times to Walenbos can be calculated using the flow tracing routine. The results are shown in Figure 5.

It can be concluded that the contributing infiltration area is almost three times as large as would be deduced from topography only. Hence, it appears that part of the precipitation that infiltrates in the area south of Walenbos, and which topographically belongs to another catchment, is contributing to the groundwater system that delivers water to the Walenbos seepage area. That such a situation is very likely, is supported by the abundant amount of seepage in Walenbos, characterising the site as a major wetland area in the region. The groundwater flow in the unconfined aquifer takes less than 100 years before it reaches Walenbos. At the northern and southern boundary of the infiltration area, groundwater flows through the confined aquifer with travel times of several hundred years before reaching Walenbos.

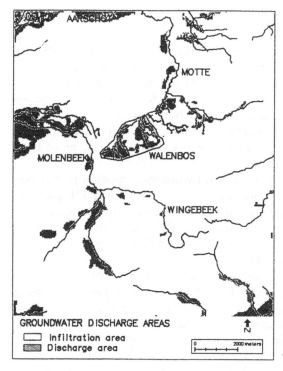

Figure 4. Calculated groundwater discharge areas, overlain by the river system.

Figure 5. Contributing area of Walenbos and calculated flow times.

3   VEGETATION AND PIEZOMETRY

Walenbos has always been an important wetland area. In the last century attempts were made to reclaim large parts of the area as meadows. However, this turned out to be in vain, since due to the wet soil, intensive agriculture and wood production with fast growing trees such as Canadian popular, Japanese larch, etc. could only succeed by maintaining a labour intensive drainage system. Consequently the Flemish Institute of Nature Conservation recently decided to restore the area to its natural state, by letting nature take its course. Today the nature reserve consists of 350 ha of forest and only 20 ha of meadows and open areas.

For the last 4 years an intensive scheme of eco-hydrological monitoring has been undertaken. Groundwater levels are recorded regularly in more than 110 piezometers. Vertical gradients in groundwater pressure are measured in double piezometers to map locations of upward groundwater discharge (Fig. 6d). Also, areas where measured seasonal groundwater level differences are small correspond with areas having upward groundwater fluxes (Fig. 6c).

Of the 70 different mapped plant species and 15 mosses several phreatophytes are indicators of groundwater discharge, De Becker (1993). Figure 6b presents some results of the vegetation mappings. The oldest forest parts with species such as beach and oak are situated along the south side on the colluvium at the foot of the valley wall. Just north of this zone appear oligotrophic and mesotrophic alder carr with a species-rich shrub, herb and moss layer, dominated by extensive peatmoss (*Sphagnum div.* spp.) carpets and many phreatophytes. In this zone high groundwater discharge in the remains of the drainage system is found. This drainage system is often blocked by colony forming iron-oxidizing bacteria (*Gallionella ferruginea*) which bind the iron from the groundwater. Further northwards in the valley, eutrophic alder

GROUNDWATER SIMULATION   **VEGETATION**

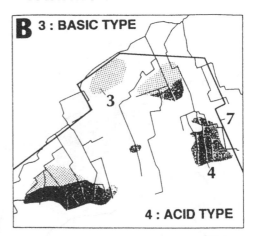

Figure 6. Simulated and measured groundwater discharge in Walenbos, a) Modelling results, (1: high discharge, 2: low discharge), b) Mapped discharge dependent vegetation (3: basic type, 4: acid type).

PIEZOMETRY IN TIME   **GROUNDWATER PRESSURE**

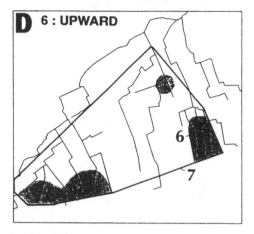

Figure 6. Continued. c) Waterlevel changes smaller than 40 cm (5), d) Areas with groundwater pressure differences indicative of upward groundwater discharge (6), and boundary of area of field measurements (7).

forests are found due to stagnant and ion-rich water. On the dryer spots appears alder-bird cherry forest. From comparison of the groundwater model results (Fig. 6a) with vegetation mappings and groundwater level measurements (Fig. 6b, c, d), it can be concluded that the vegetation and piezometric patterns give a good identification of the groundwater discharge system. Differences in the mapped patterns in Figure 6 have to be attributed to the relatively short measurement period of the groundwater levels, local inhomogeneities in the soil due to colluvial material, and slow change in vegetation when hydrological conditions change.

4 HYDROCHEMICAL ANALYSES

In the first half of 1993, two series of water samples were taken from piezometers distributed over the nature reserve and chemical analyses undertaken. The results of the analyses of the two series are very similar, so that only the last series of May 1993 will be discussed. In this series 125 samples were collected in the piezometric network of the nature reserve. Each sample was analyzed for major cations (Ca^{2+}, Mg^{2+}, Na^+, K^+, Fe^{2+}, NH_4^+), anions (HCO_3^-, $H_2PO_4^-$, NO_3^-, SO_4^{2-}, Cl^-), pH and electrical conductivity. Twenty-one percent of the samples with an absolute electrical neutrality error greater than 5% were excluded; 99 samples were retained.

Several iso-concentration mappings reveal the spatial distribution of different seepage waters. Different groundwater types can be identified by cluster analysis based on a hierarchical fusion technique in combination with Ward's minimum variance linking method (De Becker, 1993). Four clusters can be identified (De Becker, 1993). Table 1 gives the minimum, maximum, average and standard deviation of the different chemical parameters in the clusters. Figure 7 shows the spatial distribution of the clusters over Walenbos.

Table 1. Statistics for clusters of samples on 5/93.

Cluster 1 16 sampl.	Cond µS/cm	pH	HCO$_3^-$ [ppm]	H$_2$PO$_4^-$ [ppm]	NO$_3^-$ [ppm]	NH$_4^+$ [ppm]	SO$_4^{2-}$ [ppm]	Cl$^-$ [ppm]	Na$^+$ [ppm]	K$^+$ [ppm]	Ca^{2+} [ppm]	Mg^{2+} [ppm]	Fe^{2+} [ppm]
Min.	118	4.7	1.5	0.016	0.022	0.006	13	22	9.2	0.3	6	1.5	0.12
Max.	1154	6.4	71.0	1.347	37.621	6.955	359	77	33.0	6.3	145	18.0	39.00
Aver.	386	5.5	20.9	0.176	5.401	0.537	77	47	16.9	3.2	32	6.0	9.09
Std.dev.	60	0.1	4.9	0.087	2.913	0.429	20	4	1.7	0.4	8	1.1	3.18
Cluster 2, 35 samples													
Min.	218	5.7	39.0	0.016	0.022	0.006	7	16	8.4	0.8	19	1.8	0.07
Max.	549	6.6	135.0	0.940	3.320	0.927	96	63	20.0	4.0	66	9.4	29.60
Aver.	334	6.2	79.9	0.083	0.190	0.100	31	33	11.6	2.4	34	4.7	8.75
Std.dev.	15	0.0	3.7	0.028	0.094	0.028	4	2	0.6	0.1	2	0.3	1.27
Cluster 3, 9 samples													
Min.	494	6.2	140.0	0.031	0.022	0.006	93	16	8.4	1.2	74	8.3	0.15
Max.	1202	7.0	355.0	0.188	1.018	0.374	319	37	15.0	3.4	200	19.0	23.10
Aver.	805	6.7	225.0	0.059	0.413	0.140	182	28	11.4	2.1	121	13.7	8.46
Std.dev.	72	0.1	20.8	0.018	0.142	0.045	26	2	0.7	0.2	12	1.2	2.37
Cluster 4, 39 samples													
Min.	255	6.3	93.0	0.016	0.022	0.006	4	9	4.8	0.7	27	3.1	0.01
Max.	1325	7.5	625.0	0.626	2.611	1.932	185	31	17.0	5.7	230	18.0	14.40
Aver.	452	6.9	222.6	0.114	0.380	0.173	28	15	8.2	2.3	67	7.7	3.38
Std.dev.	31	0.1	16.5	0.022	0.092	0.058	5	1	0.3	0.2	6	0.6	0.55

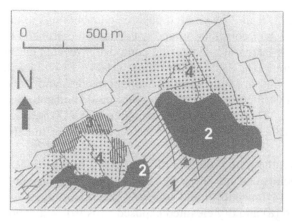

Figure 7. Spatial distribution of hydro-chemical groundwater clusters on May 1993.

1-Cluster 1   2-Cluster 2   3-Cluster 3   4-Cluster 4

From principal component analysis of the chemical parameters, it can be concluded that electrical conductivity, Ca^{2+}, Mg^{2+}, SO_4^{2-} and HCO_3^- are highly correlated and are the most influential parameters in the data set, explaining almost 60% of the total variance in the series. Cluster 1 has the lowest pH, HCO_3^- and Ca^{2+} and the highest values for Na^{2+}, Cl^-, Fe^{2+}, NO_3^-, K^+, NH_4^+ and $H_2PO_4^-$, and is clearly outstanding in chemical characteristics. It can be described as metatrophic water with precipitation characteristics. Possible effects of human activity, such as sewage water discharge and nitrification, are noticeable in some of the relatively high concentrations of the ions. The high Fe^{2+} concentration can be explained by the glauconite mineral which is abundant in the phreatic aquifer. The cluster is located at the foot of the hills south of Walenbos, in the active southern discharge zones and on the ridge between the two sub-basins in Walenbos. Cluster 2 has a quality in between clusters 1 and 4. It is located in a narrow band, north of cluster 1 in the active discharge zone. Cluster 3 is chemically similar to cluster 2, but an outstanding feature is its high SO_4^{2-} concentration. Due to the small number of samples belonging to this cluster interpretation is difficult. The cluster appears only in a belt at the northern rim of the western sub-basin of Walenbos. Cluster 4 is in quality opposite to cluster 1. It has a high concentration of Ca^{2+} and HCO_3^- and is the most alkaline cluster, probably due to seeping groundwater from the deep semi-confined aquifer. It can be described as a lithotrophic type of water.

5 REMOTE SENSING

In order to investigate the possibility of remote sensing for identification of groundwater discharge patterns, a Landsat Thematic Mapper (TM) image of 20 September 1986 was analysed. Different methods were applied, such as Normalized Difference Vegetation Index (NDVI), single band ratios, Principal Components Analysis (PCA) and tasseled cap. PCA appeared to be the most efficient data reduction method, using six TM bands and excluding the thermal band. The first component explains almost 70% of the variance, while the first and second components together explain more than 94% of the variance. TM 1, 2, 3, 5 and 7 give high positive contributions to the first component, while TM 4 has a lower contribution and is negatively correlated to the first component. Analyses of the histogram of the first component reveals that it consists of three normal distributions (Batelaan & De Smedt, 1994). Bobba et al. (1992) suggested in an analysis of Landsat Multi-Spectral Scanner bands 5 and 7 and their ratio, that the three normal distributions in the histogram could indicate discharge, transition and recharge areas. Therefore, the first component of PCA was reclassified in nine different groups (Fig. 8), making use of the eccentricity of the distributions.

Field investigation indicated that the nine groups could be related to the wetness of the areas, ranging from deep ponds and marshes (1), shallow ponds and rivulets (2), forests (3), borders of forests (4), meadows and orchards (5), urban areas and roads (6 and 7), cultivated farmland (8) and dry elevated exposed soils (9). Walenbos appears as a very wet area; hence, a typical seepage area. Also, several other groundwater discharge areas could be identified.

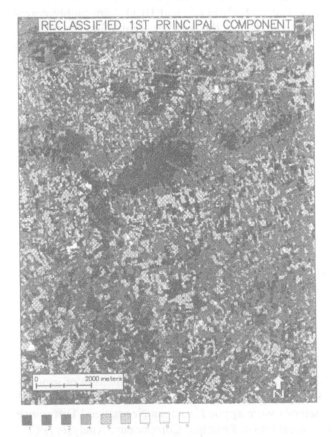

Figure 8. Reclassified first principal component of the Landsat Thematic Mapper image, resulting in a classified map of wet to dry areas, indicated by grey values 1-9.

6 INTEGRATION OF RESULTS AND CONCLUSIONS

By integrating the different results of the groundwater discharge analyses, new insights into the relationships between groundwater and the environment can be obtained.

Remote sensing analyses yield a regional picture of the groundwater discharge situation. Comparing the wettest areas of the reclassified principal component 1 with the simulated groundwater discharge areas results in a coincidence of only 16.1%. Many infiltration areas are classified as wet areas in the remote sensing analyses because of waterlogging and ponding. Therefore, at present remote sensing can only be used as a partial indicator of groundwater discharge areas.

The vegetation mapping, together with the chemical groundwater analyses, clearly give a more precise and complete description of the groundwater discharge system. For instance, the vegetation in the area of cluster 1 (Fig. 7), at the foot of the hill south of Walenbos, is dominated by oligotrophic alder forest. This vegetation type is dependent on the occurrence of groundwater discharge of acid water with low alkalinity, high iron content and small contemporary pollutions. This quality and the high seepage rate, as simulated by the model, indicate that this area is dominated by discharge from the unconfined aquifer with relatively short transport times. The high iron content of the discharge water can be explained by the abundant presence of glauconite in the phreatic aquifer formation.

In the lowest part of the valley, cluster 4 defines a less acid, more alkaline type with high HCO_3^- Ca^{2+} and low Fe^{2+}, Cl^-, Na^+ and SO_4^{2-} concentrations. Clearly, this water is of lithotrophic origin, coming from the semi-confined aquifer. The fine sandstone formation of Brussels, in which this semi-confined aquifer is situated, is known to have some calcium carbonate in its matrix. Hence, the groundwater flow system belonging to the area of this cluster is characterized by long flow lines, discharging in the lowest parts of the valley and a lithologically determined quality. Also, the quality of the water and the fact that the discharged groundwater is stagnant in this zone, causes the vegetation to be meso- or eutrophic, i.e. mainly alder forests.

Groundwater modelling is clearly the most useful technique for identifying seepage zones on a regional scale. For ecohydrological identification purposes, as demonstrated here, the technique is sufficiently accurate. Local groundwater level measurements can support the modelling results, but can also point to deviations in modelled results due to incomplete hydrogeological description and/or temporal variations. The groundwater quality analyses are very useful in differentiating between seepage zones. In combination with the modelled results they are very interesting and assist in understanding the link between the groundwater system and vegetation. By using both quantitative and qualitative information on the seepage areas, the physical characteristics and requirements of the seepage vegetation types can be well understood.

Hence, this study clearly shows that when the results of the different approaches are combined, a more detailed and complete understanding of the groundwater system and how this impacts on the landscape and its ecological features, is obtained.

ACKNOWLEDGMENT

This research was supported by a grant from the National Fund for Scientific Research of Belgium.

REFERENCES

Anderson, M.P. & Woessner, W.W. 1992. *Applied Groundwater Modelling – Simulation of Flow and Advective Transport. San Diego: Academic Press Inc.*

Batelaan, O., De Smedt, F., Otero Valle, M.N. & Huybrechts, W. 1993. Development and application of a groundwater model integrated in the GIS Grass. In Kovar, K. & Nachtnebel, H.P., *Application of Geographic Information Systems in Hydrology and Water Resources Management. IAHS publ.* 211.

Batelaan, O. & De Smedt, F. 1994. Use of Landsat TM in the analysis of groundwater flow systems. In: *Space Scientific research in Belgium,* vol III, Earth Observation Part 2, pp. 215-233, Federal Office for Scientific, Technical and Cultural Affairs.

Bobba, A.G., Bukata, R.P. & Jerome, J.H. 1992. Digitally processed satellite data as a tool in detecting potential groundwater flow systems. *J. Hydrol.* 131: 25-62.

Bronders, J. 1989. Contribution to the geohydrology of Mid-Belgium by means of geostatistical analysis and a numerical model (in Dutch). Ph.D. thesis, Free University Brussels, Belgium.

De Becker, P. 1993. Chemical characteristics of the groundwater in the state nature reserve 'Het Walenbos' (in Dutch*). M.Sc. thesis, Free University Brussels, Belgium.*

Freeze, R.A. & Cherry, J.A. 1979. Groundwater. Prentice-Hall Inc., *Englewood Cliffs, N.J.*

Hacksbuch, W. 1985. *Multi-grid Methods and Applications.* Springer-Verlag. Berlin.

Strack, O.D.L. 1984. Three-dimensional streamlines in Dupuit-Forchheimer models. *Water Resources Res.* 20(7): 812-822.

CHAPTER 8

Comparison of standardized and region-specific methods for assessment of the vulnerability of groundwater to pollution: A case study in an agricultural catchment

C. BARBER, L.E. BATES, R. BARRON & H. ALLISON
CSIRO Division of Water Resources, Private Bag, Australia

ABSTRACT: The DRASTIC vulnerability assessment procedure, a standardized system developed in the US, has been evaluated against a simpler, region-specific spatial modelling approach in an agricultural area of complex geology in northern New South Wales. There was a broad relationship between nitrate in groundwater and relative vulnerability determined by both DRASTIC and spatial modelling. Both assessments, however, underestimated vulnerability to pollution by nitrate particularly for point sources such as poultry farms and piggeries. Vulnerability maps derived from spatial models are much easier to develop and apply than DRASTIC, and overcome the subjectivity of the standardized approach. They still underestimate vulnerability to pollutants like nitrate which behave semi-conservatively in soils and groundwaters in the study area. This emphasizes the need for caution in applying vulnerability maps, which are only meant to be regional scale guides and should not be used to infer pollutant behaviour at local scale.

1 INTRODUCTION

A variety of techniques for vulnerability assessment have been developed over the last decade to help provide guidance to developers, planning authorities and regulatory agencies in relation to groundwater protection. The DRASTIC system of vulnerability assessment is widely used in the US, although the few attempts to validate this approach have met with a mixed response e.g. USEPA (1992), Barber et al. (1993) and Kalinski et al. (1994). DRASTIC has been criticised because of its subjectivity, and because of its large data requirements (see Meeks & Dean, 1990 and Barber et al., 1993).

It has been made very clear in Australia that some form of relative vulnerability assessment at regional scale is required. However, there is concern that these assessments will be used too prescriptively, and ultimately could become weak but inexpensive replacements for groundwater and environmental impact assessments (EIA's). Vulnerability assessments like DRASTIC were designed only as a guide to likely vulnerability, and were not intended to (and cannot) be used as a replacement for EIA's.

To overcome some of the drawbacks of the standardized approach to regional vulnerability assessment, we have investigated spatial modelling techniques to establish area (basin) specific relationships between physical variables and actual or inferred measures of contamination of groundwater. The aims of the project were:

1. To identify and use a small number of key predictor variables within the model, to simplify application;

2. To calibrate the spatial model in other similar groundwater basins to overcome the subjectivity of the standardized approach;

3. To provide mechanisms for quantifying uncertainty of vulnerability, in an effort to avoid prescriptive use of the generalised approach.

The model is being applied in three catchments of varying geological complexity. These are the Peel catchment in northern New South Wales, an area in the southeast of South Australia dominated by Gambier Limestone, and the Swan Coastal Plain in Western Australia. The results of the study comparing vulnerability defined by spatial modelling in the Peel catchment with earlier assessments using DRASTIC by Barber et al. (1994) are reported here.

2 STUDY AREA

The study area is situated around the rural township of Tamworth in northern New South Wales (see Fig. 1). Principle land uses are grazing, dryland and irrigated (lucerne) cropping, and more recently intensive animal husbandry (mostly piggeries and poultry farms).

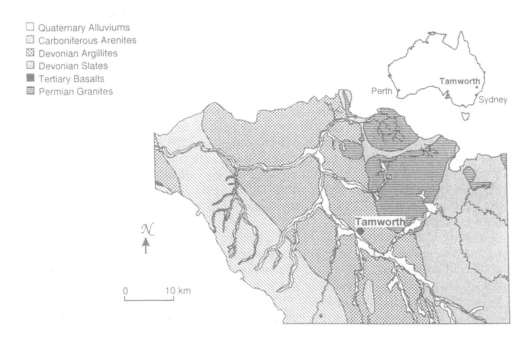

□ Quaternary Alluviums
▧ Carboniferous Arenites
▨ Devonian Argillites
▨ Devonian Slates
▪ Tertiary Basalts
▤ Permian Granites

Figure 1. Generalized geology of the Peel Catchment, northern New South Wales.

The geology of the area as shown in Figure 1 is quite complex. There are three main aquifers in the region. Alluvial gravel/sand/clay sequences within fluvial palaeochannels of limited lateral extent constitute the highest yielding aquifer with yields up to 20 L s^{-1}. Other aquifer systems consist of more extensive fractured Carboniferous and Devonian interbedded shales, sandstones and occasional limestones, and fractured Permian granites with yields of less than 1 L s^{-1}. The alluvials have been exploited for town supplies and irrigation while the fractured rock aquifers are used extensively for stock watering and rural supplies.

3 STANDARDIZED REGIONAL VULNERABILITY ASSESSMENT

The standardized DRASTIC system of Aller et al. (1987) was used to define vulnerability initially, and details of this are given in Barber et al. (1994). DRASTIC is an acronym defining the spatial variables used to assess vulnerability. These are Depth to groundwater, net Recharge, Aquifer media, Soil type, Topography (slope), Impact of vadose zone and hydraulic Conductivity. Depth to groundwater and vadose zone impact are the most highly weighted variables.

The DRASTIC vulnerability map is shown in Figure 2, along with summary data for nitrate in groundwater for expected low-nitrogen source land-uses (nitrate less than or greater than 1.5 mg L^{-1}) and point sources (nitrate above or below 10 mg L^{-1}). The latter data were collected as part of a research program evaluating DRASTIC, described in Barber et al. (1994).

In general, the western part of the catchment is of lower vulnerability, as defined by DRASTIC. The alluvial deposits and granites in the eastern part are relatively more

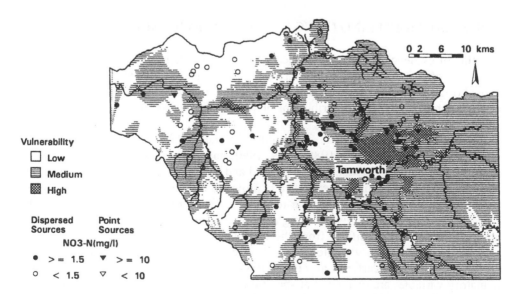

Figure 2. DRASTIC vulnerability map of the Peel catchment, showing nitrate occurrences in groundwater for low nitrogen loading dispersed sources, and high loading point sources (poultry farms and piggeries).

Table 1. Nitrate concentrations in groundwater from wells near point sources (piggeries, poultry farms) in relation to relative vulnerability defined by DRASTIC and a weights of evidence spatial model.

	Vulnerability		
	Low	Medium	High
DRASTIC			
No. bores sampled	8	12	7
NO_3-N > 10 mg L^{-1}	6	5	7
Range NO_3-N (mg L^{-1})	5.5-42.5	0.3-52.0	10.5-58.0
Spatial Model			
No. bores sampled	13	1	13
NO_3-N > 10 mg L^{-1}	8	1	9
Range NO_3-N (mg L^{-1})	5.5-42.5	13.0	0.3-58.0

vulnerable, and show a greater incidence of nitrate-N concentrations in groundwater greater than 1.5 mg L^{-1} beneath low-nitrogen landuses. As argued in Barber et al. (1994), this provides some validity to the DRASTIC approach in determination of relative vulnerability.

However, data for point sources are much less convincing and even in the low vulnerability category, the majority of these show nitrate in concentrations over the drinking water standard of 10 mg L^{-1} (Table 1). The latter indicates that, in absolute terms, DRASTIC grossly underestimates vulnerability to pollution by nitrate in the study area.

4   SPATIAL MODELLING USING WEIGHTS OF EVIDENCE

Weights of evidence is a statistical method of pattern integration based on Bayesian principles of conditional probability. The term 'weights of evidence' reflects the statistical origin of the approach which was devised to quantify the strength or power of an explanatory variable in describing or predicting the occurrence of a response variable.

To date the most notable use of weights of evidence has been in the geosciences where Bonham-Carter et al. (1988, 1989, 1990) have used weights of evidence with spatial data to predict the occurrence of gold and other mineral deposits. When dealing with mapped information, evidence is provided by the unique spatial association of each binary map pattern with the response variable of interest, for example observed gold deposits. The use of weights of evidence to assess the regional vulnerability of groundwater differs from these applications as the occurrence of the response variable is not 'weighted' by map area as is the case in Agterberg et al. (1990) and Bonham-Carter et al. (1988, 1989, 1990). Instead, the response and explanatory variables are evaluated at specific points throughout the study area to provide inputs to the analysis in the form of frequencies. In vulnerability mapping hydrogeological maps and point data defining groundwater quality are used as data inputs.

The strength of the association between input maps and the response variable is expressed by a pair of coefficients, the positive and negative weights of evidence. A separate pair of weights is calculated for each explanatory input map considered. Prior knowledge of the response variable is used in conjunction with the weights to derive posterior probability indices. These are used to predict further occurrence of the response variable for each unique combination of explanatory factors. As vulnerability cannot be measured directly, it is necessary to rely on indicators which either record current levels of contamination or characterize the chemical nature of groundwater at specific locations. Both the explanatory and response variables are binary in nature. Explanatory map inputs are reclassified to give a binary map pattern and a threshold value is chosen for the response variable so that observations are classified as either 'vulnerable' or 'not vulnerable'.

4.1 *Mathematical model*

The mathematical model for weights of evidence can be described as follows. The prior probability that groundwater at a particular location is vulnerable is given by:

$$P_{\text{prior}} \{V\} = \frac{N\{V\}}{N\{T\}} \tag{1}$$

where $N\{T\}$ is the total number of observations and $N\{T\}$ is the number of observations assessed as 'vulnerable'.

This is assumed to be constant for the study area. The prior probability is transformed to the prior odds as follows:

$$O_{\text{prior}} \{V\} = \frac{P_{\text{prior}} \{V\}}{1 - P_{\text{prior}} \{V\}} \tag{2}$$

The conditional probability that a location will be 'vulnerable' (V) given the explanatory map pattern D_1 is:

$$P_{\text{post}} \{V|D_1\} = \frac{P_{\text{post}} \{D_1|V\} \, P_{\text{prior}} \{V\}}{P_{\text{prior}} \{D_1\}} \tag{3}$$

Similarly, the conditional probability that a location will not be 'vulnerable' ($\bar{V}$) given the explanatory map pattern D_1 is:

$$P_{\text{post}} \{\bar{V}|D_1\} = \frac{P_{\text{post}} \{D_1|\bar{V}\} \, P_{\text{prior}} \{\bar{V}\}}{P_{\text{prior}} \{D_1\}} \tag{4}$$

A further two conditional probabilities can be calculated for those cases where the explanatory map pattern D_1 is absent. The calculation of the posterior probability is based on 'logits' (the natural logarithm of odds ratio) where:

$$\log_e O_{\text{post}} \{V|D_1\} = W_{D_1}^+ + \log_e O_{\text{prior}} \{V\} \tag{5}$$

$$\log_e O_{\text{post}} \{V|\overline{D}_1\} = W_{\overline{D}_1}^- + \log_e P_{\text{prior}} \{V\} \tag{6}$$

where $W_{D_1}^+$ and $W_{D_1}^-$ are the positive and negative weights of evidence:

$$W_{D_1}^+ = \log_e \frac{P_{\text{post}} \{D_1|V\}}{P_{\text{post}} \{D_1|\overline{V}\}} \tag{7}$$

$$W_{D_1}^- = \log_e \frac{P_{\text{post}} \{\overline{D}_1|V\}}{P_{\text{post}} \{\overline{D}_1|\overline{V}\}} \tag{8}$$

The positive weight is used in the presence of the binary map pattern while the negative weight is used in its absence. The posterior probability that a location is vulnerable given the map pattern is derived from Equations (5) and (6) using the relationship between odds and probability:

$$P_{\text{post}} \{V|D_1\} = \frac{O_{\text{post}} \{V|D_1\}}{1 + O_{\text{post}} \{V|D_1\}} \tag{9}$$

Evidence from a number of mapped hydrogeological factors is incorporated by employing Bayes Rule for conditional probability, such that the posterior odds that a location will be vulnerable given a series of binary predictor patterns $(D_1, D_2 \dots D_j)$ is:

$$\log_e O_{\text{post}} \{V|D_1 D_2 \dots D_j\} = W_{D_1}^+ + W_{D_2}^+ + \dots W_{D_j}^+ + \log_e O_{\text{prior}} \{V\} \tag{10}$$

Each combination of predictor patterns can be thought of as a unique condition. A matrix of probability indices for each unique condition is derived by extending Equation (9) as follows:

$$P_{\text{post}} \{V|D_1 D_2 \dots D_j\} = \frac{O_{\text{post}} \{V|D_1 D_2 \dots D_j\}}{1 + O_{\text{post}} \{V|D_1 D_2 \dots D_j\}} \tag{11}$$

This expression allows vulnerability to be defined spatially, if a large enough number of relationships between response variable (V) and predictors $(D_1 D_2 \dots D_j)$ are known, and if the spatial patterns (map patterns) of suitable predictors are known. Thus weights of evidence provides a mechanism for deriving a groundwater vulnerability map from known relationships.

5   SPATIAL MODELLING OF VULNERABILITY

In using weights of evidence to derive probability indices for groundwater vulnerability mapping, three easily defined, key predictor variables were selected for their importance in determining residence time in the vadose zone and shallow groundwater. These were depth to groundwater, vadose zone impact as defined for the catchment by Barber et al. (1994), and lithology (alluvial, fractured interbedded sedimentary sequences and fractured granite). Other predictor variables such as those in the DRASTIC system were discarded as being of much lesser importance in determination of vulnerability. Lithology was used also to provide a close association with physical geology.

Predictor variables were split further into three binary dummy variables as shown in Table 2 (depth: 0-5, 5-15 and >15 m; vadose scores: 1 to 3, 4 to 5 and >5; lithology: alluvial, fractured interbedded sedimentary sequences and fractured granite). Thus we used nine predictor variables, analysing the occurrence of each in relation to the response variable, giving 27 predictor patterns. As described above, the weights of evidence are used to calculate the posterior probability indices used to depict spatial variations in groundwater vulnerability across the study area. Generally, when using weights of evidence analysis, the negative weight is applied in the absence of the predictor variable (see Eq. 6). However, in our application each major class of predictor variable, depth to groundwater, vadose zone character and lithology, is further broken down into three mutually exclusive dummy variables. Therefore the negative weights of evidence is not used.

Choice of response variable was more problematic. Initially we considered nitrate in groundwater for each of over 150 boreholes sampled in the area as part of the evaluation of DRASTIC. In this way, we used point data (response and predictor variables) to establish the overall relationship between predictor pattern and response to calibrate the model. A Geographic Information System (ARC/INFO) was then used to map each of the 27 predictor patterns, to assign the calibrated probability indices to these and in so doing determine relative vulnerability. For convenience, vulnerability was determined as low, medium and high.

Results using nitrate as response for low nitrogen loading land uses proved unsatisfactory in producing a useable vulnerability pattern. In general, intermediate depths to groundwater were rated most vulnerable, possibly because of variable denitrification in areas with shallow depth to groundwater which consequently had lowest vulnerability (to nitrate). This dominated the pattern and juxtaposed high and low vulnerabilities in close proximity near areas with shallow depth to groundwater.

As an alternative we used natural levels of electrical conductivity (EC) in shallow groundwater (i.e. where EC is not unduly affected by anthropogenic sources of contamination) as a surrogate for residence time in the unsaturated zone. Thus where resi-

Table 2. Predictor variables used in weights of evidence analysis of regional groundwater vulnerability.

Predictor variables	Variate	DRASTIC rating	Description
Depth to groundwater	d_1	1-3	>15 m
	d_2	5-7	5-15 m
	d_3	9-10	< 5 m
Vadose zone	v_1	1-3	
	v_2	4-5	
	v_3	6-8	
Lithology	l_1		Fractured sedimentary
	l_2		Alluvials
	l_3		Fractured granites

dence time is low (rapid transit), EC would tend to be lower as there is likely to be lower levels of soil carbon dioxide in pore water and less weathering of minerals. Wendland et al. (1994) makes use of this proposition in his study of regional vulnerability to nitrate contamination in Germany. Also, Lawrence (1983) has previously determined that elevated nitrate concentrations occur where salinities are low because of high recharge rates for a basalt aquifer in western Victoria. Further work by Lawrence (1983b) has shown that this is generally true throughout Australia. A relationship between low EC and high DRASTIC vulnerability has already been shown to be statistically significant by Barber et al. (1994) for the Peel catchment. For convenience, a binary response variable of > or < 1500 mS cm^{-1} for EC was used. Also, because of the relatively small sample of data available for calibration, it became necessary to assign a nominal initial value of 0.5 to each predictor pattern to avoid recording zero occurrences of pattern which made probability calculations impossible.

The resulting probabilities for each of the 27 predictor patterns are shown in Table 3. Figure 3 depicts the groundwater vulnerability map derived by assigning the probability indices in Table 3 to each unique condition. It is clear that for each lithology, relative vulnerability increases as depth to groundwater decreases, and as vadose zone score increases (decreasing clay content and locally unconfined conditions). Additionally, shales are least vulnerable whilst granites are most vulnerable, closely followed by alluvial deposits. Probability indices are not assigned to those 'unique conditions' which in reality do not occur in the study area; for example, granitic areas with shallow depth to groundwater.

The determined relative vulnerabilities show a similar pattern to DRASTIC. The effort put into obtaining spatial coverages for three variables is much less than that needed to determine the seven for the DRASTIC map. Additionally, the simple sampling program of available bores in the calibration area for determination of EC precludes the subjective score assignment process inherent in the DRASTIC system.

The main drawbacks to the spatial modelling approach are the uncertainties in the determined probabilities which are based of necessity on very few (150) samples. Although increasing the sample number from an initial 100 carried out in the original study of Barber et al. (1994) to 150 did not change probabilities significantly, a much larger sample size would be preferable to allow assessment of the significance of the

Table 3. Relative probability indices for the Peel River catchment.

Lithology	Vadose zone scores			Depth to groundwater (m)
	>5	4-5	<4	
Granites	.99	.95	.93	>15
	.99	.97	–	5-15
	.99	.98	–	< 5
Alluvials	.91	.55	.45	>15
	.94	.66	.56	5-15
	.96	.77	.69	< 5
Fractured sedimentary	.81	.35	.26	>15
	.87	.46	.36	5-15
	.92	.59	.49	< 5

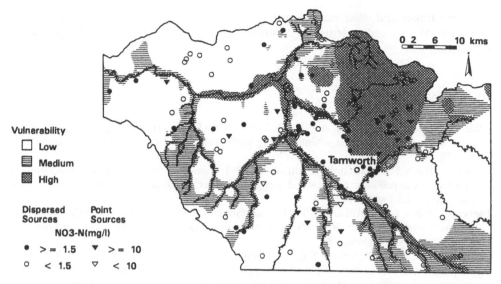

Figure 3. Vulnerability map derived using weights of evidence spatial modelling, showing nitrate occurrences in groundwater as in Figure 2.

approach. At present, these can only be regarded as empirically derived relationships.

It is also clear that spatial modelling determines universal vulnerability, as does DRASTIC. Consequently, in absolute terms the assignment of low medium and high vulnerabilities could convey the wrong message for contaminants like nitrate which appear to behave semi-conservatively over most of the catchment (see Table 1). In the case of nitrate, all groundwater is vulnerable in the study area and assessments should take this into account. Further work is being carried out to quantify these uncertainties.

6 SUMMARY

The standardized DRASTIC system and spatial modelling produced similar relative vulnerability maps, although both underestimated vulnerability to nitrate which behaves semi-conservatively in soil and groundwater in the study area.

Spatial models using weights of evidence techniques and calibrated using natural groundwater EC were much easier to use than DRASTIC, and because they were calibrated they overcame the subjectivity inherent in the DRASTIC approach.

There is a need to define uncertainties in vulnerability assessments to avoid overly prescriptive use of vulnerability maps. Results here suggest that vulnerability maps should be used with caution. At best these are only regional scale guides, and cannot be used to infer contaminant behaviour or impact at local scale.

ACKNOWLEDGEMENTS

The authors would like to thank George Gates, Michael Williams, Dan McKibbin,

Matthew Baker and other staff from the NSW Department of Water Resources for provision of data and their useful contributions to discussions during the course of this study.

REFERENCES

Aller, L., Bennett, T., Lehr, J.H. & Petty, R. J. 1987. DRASTIC, a standardized system for evaluating groundwater pollution potential using hydrogeological settings. USEPA Report 600/2-87/035.

Barber, C., Bates, L.E., Barron, R. & Allison, H. 1993. Assessment of the relative vulnerability of groundwater to pollution: a review and background paper. *AGSO Journal of Australian Geology and Geophysics* 14(2/3): 147-154.

Barber, C., Bates, L.E. & Allison, H. 1994. Evaluation of DRASTIC – a regional vulnerability assessment procedure. *Proc. International Symposium on Groundwater, Drought, Pollution and Management*. Wallingford, UK.

Bonham-Carter, G.F. & Agterberg, F.P. 1990. Application of a microcomputer-based. Geographic Information System to mineral potential mapping. In: J.T. Hanley & D.F. Merriam (eds), *Microcomputer Applications in Geology Volume 2*, pp 49-74. Oxford, Pergamon Press.

Bonham-Carter, G.F., Agterberg, F.P. & Wright, D.F. 1988. Integration of geological datasets for gold exploration in Nova Scotia. *Photogrammetric Engineering and Remote Sensing* 54(11): 1585-1592

Bonham-Carter, G.F., Agterberg, F.P. & Wright, D.F. 1989. Weights of evidence modelling: a new approach to mapping mineral potential. In: F.P. Agterberg & G.F. Bonham-Carter (eds), *Statistical Applications in the Earth Sciences*. Geological Survey of Canada, Paper 89-9: 171-183.

Bonham-Carter, G.F., Agterberg, F.P. & Wright, D.F. 1990. Statistical pattern integration for mineral exploration. In: G. Gaal & D.F. Merriam (eds), *Computer Applications in Resource Estimation: Predictions and Assessment for Metals and Petroleum*, pp 1-23. Oxford, Pergamon Press.

Kalinski, R.J., Kelly, W.E., Bogardi, I., Ehrman, R.L. & Yamamoto, P.D. 1994. Correlation between DRASTIC vulnerabilities and incidents of VOC contamination of municipal wells in Nebraska. *Groundwater* 32(1): 31-34.

Lawrence, C.R. 1983. Nitrate rich groundwater of Australia. *AWRC Technical Paper 79*. Canberra, Australian Government Printing Service.

Lawrence, C.R. 1983b. Occurrence and genesis of nitrate-rich groundwaters of Australia. *Proc. International Conference on Groundwater and Man*, Sydney, Australia.

Meeks, Y.J. & Dean, J.D. 1990. Evaluating groundwater vulnerability to pesticides. *Journal Water Resources Planning and Management* 116(5): 693-707.

USEPA. 1992. National pesticide survey: update and summary of phase 2 results. EPA Report 590/9-91-021.

Wendland, F., Albert, H., Bach, M. & Schmidt, R. 1994. Potential nitrate pollution of groundwater in Germany: A supraregional differentiated model. *Environmental Geology* 24: 1-6.

CHAPTER 9

The dynamics of groundwater flow in the regolith of Uganda

RICHARD G. TAYLOR & KEN W.F. HOWARD
Groundwater Research Group, University of Toronto, Scarborough Campus, Ontario, Canada

ABSTRACT: Groundwater resources are developed throughout equatorial Africa in an effort to provide largely rural populations with a source of potable water. Over the last decade, interest in the shallow groundwaters of the regolith has gained considerable momentum with the recognition that the regolith may provide a more sustainable and less costly source for rural water supplies than the underlying bedrock fractures which have traditionally been exploited. Despite this attention, basic questions regarding both the geochemical evolution and the hydrogeological nature of the regolith remain unresolved. Particular concerns are the hydrogeological characteristics of the aquifer material, the hydraulic interaction of the regolith with the underlying bedrock aquifer and the nature of groundwater recharge.

A combination of piezometer tests, particle-size analysis, pumping tests, hydrochemistry, as well as regional meteorological and land use records has been used in two catchments of Uganda to investigate groundwater movement in the regolith. Results indicate that regolith aquifers are most likely to be formed in areas featuring both high rainfall and a low relief. Observed regolith profiles show strong vertical and lateral heterogeneity with aquifers at the base of the regolith possessing hydraulic conductivities between 0.3 and 3m day^{-1}. Pumping tests demonstrate that leaky aquifer conditions prevail in 80% of bedrock wells. The implied hydraulic connection between the bedrock and regolith is confirmed by drawdowns observed in adjacent, regolith observation wells during prolonged pumping of bedrock wells and stable isotope evidence that links both aquifers to the same recharge source. Recharge to the regolith-bedrock aquifer system results from the direct infiltration of rainfall during intense storms in the rainy seasons, yet estimates vary considerably from 17 to 200 mm year^{-1} between the two catchments.

1 INTRODUCTION

Groundwater drawn from fractured crystalline basement rocks of Precambrian to Palaeozoic age currently represents the primary source of potable water for the largely rural population in Uganda. Since the 1930s, over ten thousand boreholes (bedrock wells) have been put into production. Until very recently, the preferred method of well construction has been to drill wells that fully penetrate the overlying weathered mantle (regolith), and rely strictly on fractures in the competent underlying bedrock to provide an adequate well yield for handpumps. During the 1980s the high incidence of well failure in Uganda raised concerns for the long-term viability of the bedrock source, particularly as no quantitative hydrogeological investigations had been conducted prior to aquifer development. A study was initiated in 1987 (Hydrogeology-Uganda Phase I) to evaluate the resource and establish the hydrogeological nature of the bedrock aquifer.

The Phase I study was conducted in the Nyabisheki catchment in southwestern Uganda (Fig. 1), and completed in 1989 (Hydrogeology-Uganda Phase I, 1989). The results of this work strongly questioned the reliability of the bedrock aquifer for rural water supplies. Pumping tests and packer testing of boreholes showed the bedrock to be both poorly transmissive, usually less than 1 m^2 day^{-1}, and possessing a very limited storage capacity (Howard et al., 1992). Moreover, water balance calculations drew attention to low rates of recharge (17 mm $year^{-1}$) and indicated that the aquifer within the regolith provided the likely key to future resource development (Howard & Karundu, 1992; Howard, 1991). A follow-up investigation was initiated in early 1991 (Hydrogeology-Uganda Phase II) in an attempt to discern whether the hydrogeological findings in southwestern Uganda could be extrapolated to the country as a whole.

Specific objectives of this project were:

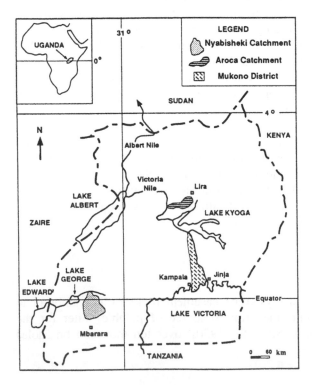

Figure 1. Location of the study areas.

1. To establish the hydrogeological characteristics of the regolith unit above the hard rock aquifer;

2. To establish the degree of hydraulic interaction between the regolith and the hard rock aquifer;

3. To determine the nature of recharge to the regolith and hard rock aquifers.

As part of this study, research activities were extended to the Aroca catchment in central Uganda (Fig. 1). A summary of the research findings is presented here.

2 STUDY AREAS

2.1 *Nyabisheki catchment*

The Nyabisheki catchment has an area of 2750 km^2 and encompasses the basin of the River Nyabisheki which runs centrally along the length of the catchment, and its tributary, the River Oruyubu. The Nyabisheki flows northward and eventually drains to Lake George, which in turn discharges into Lake Edward in the western rift. Numerous outcrops across the catchment expose basement rocks comprised primarily of granites, schists, gneisses, phyllites and quartzites of Precambrian to Palaeozoic age. Fractures occur locally and are believed to result from both regional tectonic activity associated with the western arm of the Rift Valley, and to pressure release resulting from erosion and weathering of the bedrock. Phase I studies showed that the fracture zone extends to a depth of 40 m below the bedrock surface (Howard et al., 1992).

In most areas the bedrock is concealed by a regolith, the evolution of which has been proposed by Ollier (1959). Weathering mechanisms, particularly the geochemical aspects of etchplanation, are disputed and are the subject of continuing research. However, in describing the origin of laterites, Ollier (1994) contends that the presumption of vertical processes (i.e. that regoliths represent the progressive chemical alteration of parent bedrock to a lateritic soil cover) often noted in geochemical (e.g. Nahon & Tardy, 1992) and hydrogeological (e.g. Acworth, 1987; Jones, 1985) studies is unfounded. Despite this controversy, drilling logs show that, generally, a ferricrete crust is underlain by a clay unit likely to be derived from the hydrolysis of parent rock fragments. The water table is commonly encountered at the base of the clay, below which sand-sized grains predominate to the bedrock surface. Hydrogeological studies of the regolith in Malawi (Chilton & Smith-Carington, 1984) and Nigeria (Hazell et al., 1992) have all asserted that it is within this coarse basal unit that the regolith's primary aquifer is located.

In the Nyabisheki catchment the thickness of the regolith, gauged from records of borehole casing depths (borehole casing normally extends to the base of the collapsible regolith), varies from 3 to 62 m. Well siting will have introduced some bias in these data since outcrops will have been avoided and alluvium will tend to have been preferred on account of more favourable drainage characteristics. Nevertheless, thinner and less-saturated regolith profiles were observed on hilltops and along hillslopes during the installation of regolith piezometers (Fig. 2). Such profiles can be attributed to the catchment's high relief which promotes the erosion of weathered products as valley infill and reduces infiltration of the principal weathering agent, rainfall-fed recharge. Through a statistical study of the regolith overlying the crystalline base-

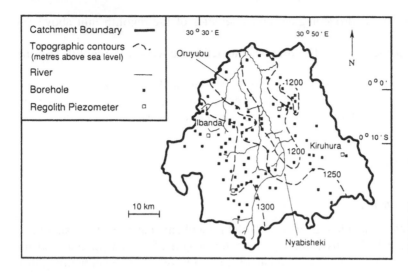

Figure 2. The Ny-
abisheki catchment
in southwestern
Uganda.

ment in Malawi, McFarlane et al. (1992) noted similarly that saturated regolith thickness decreased with increasing relief.

The amount of rainfall across the Nyabisheki catchment tends to vary with surface elevation. Approximately 1400 mm year^{-1} occurs in the higher altitudes near Ibanda (1400 metres above sea level (masl)) in the west, while roughly 750 mm year^{-1} has been recorded at an altitude of 1200 masl in the east around Kiruhura. The south-western portion of the catchment which features a combination of higher rainfall and low relief (Fig. 2) appears, based upon piezometric data and a limited number of electromagnetic surveys (Hydrogeology-Uganda Phase II, 1994), to be the only region capable of maintaining a saturated thickness of greater than 5 m within the regolith.

2.2 *Aroca catchment*

The Aroca catchment lies within the Victoria Nile basin of central Uganda and covers an area of 840 km^2. Although the underlying bedrock has not been mapped, a limited number of outcrops outside the catchment expose biotite and granulitic gneisses, characteristic of the Precambrian Aruan complex that spans much of northern Uganda (Leggo, 1974). Preliminary mineralogical data from well cuttings obtained during the installation of regolith piezometers also point to the existence of localised quartzite cover formations. Packer testing of catchment boreholes (Fig. 3) shows the bedrock is fractured to a depth of at least 30 m below the bedrock surface and, similar to the Nyabisheki catchment, provides a weakly transmissive ($\gg$ 1 m^2 day^{-1}) aquifer (Hydrogeology-Uganda Phase II, 1994).

The region receives an average rainfall of 1400 mm year^{-1} and is drained by an extensive (7% of the catchment area) marsh, sometimes referred to as a dambo, which slopes gently from 1100 masl in the northeast to 1030 masl in the southwest where it meets the Victoria Nile. In contrast to the Nyabisheki catchment, streamflow discharge in the dambo is ephemeral, occurring sporadically at the end of each rainy season in June and October. Throughout the Aroca catchment, the bedrock is entirely

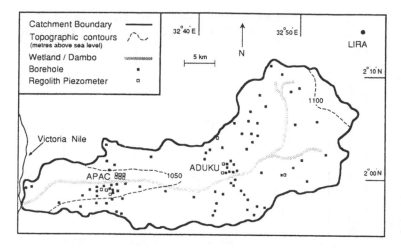

Figure 3. The Aroca catchment in central Uganda.

concealed by a regolith having a thickness that ranges from 12 to 41 m (based on borehole casing depths). The prevalence of groundwater encountered during the installation of regolith piezometers (Fig. 3) suggests that the regolith aquifer in the Aroca catchment is, unlike the Nyabisheki, regionally extensive. Such observations compare favourably with the relationships established in areas of low relief in Malawi between saturated thickness of the regolith and factors such as waning incision indicated by dambo coverage or reduced stream presence (McFarlane et al., 1992).

3 REGOLITH PROPERTIES

To investigate the hydrogeological properties of the regolith, both particle size analysis and piezometer tests were used. Particle size analyses were conducted on regolith samples collected during the construction of regolith piezometers and production wells in the bedrock. Slug and bail tests were performed in constructed piezometers. The first approach provided both a qualitative indication of permeability characteristics down the regolith profile as well as an empirically-based measure of the aquifer's hydraulic conductivity. The latter revealed the gross response of the regolith's saturated thickness to hydraulic stress. Results from piezometer tests and preliminary findings of the particle-size work are summarised here.

3.1 *Particle-size analysis*

Particle-size analysis was conducted on 42 samples from 7 regolith piezometers using standard wet-sieving procedures. The general stratigraphy, supported by particle-size profiles for four sites, is given in Figure 4. A sandy lateritic crust of iron and aluminum oxides rests over a muddy (i.e. silt and clay) zone of kaolinite, vermiculite and limonite. Below this, a sandy and typically-saturated unit of bedrock fragments featuring feldspar and mica clasts, persists to the bedrock surface. To a large extent, samples were poorly-sorted and contained quartz fragments throughout the depth of the profile. In the Aroca catchment, quartz cobbles were occasionally observed in

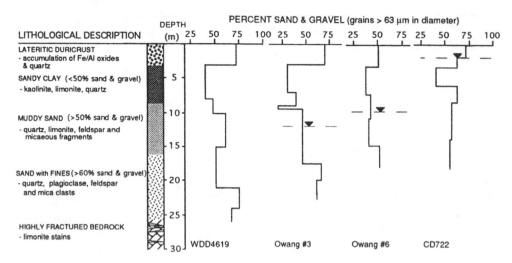

Figure 4. Regolith stratigraphy in the Aroca catchment of Uganda (supported by preliminary results from grain size analyses).

collected well cuttings and indicated the existence of presently unmapped, quartzite cover formations.

In unconsolidated media, particle size is the key factor controlling permeability. As such, the median size (d_{50}) forms the basis of the determination of intrinsic permeability, $k = c(d_{50})^2$ (Freeze & Cherry, 1979). As indicated by the fraction of sand and gravel determined in regolith profiles from the Aroca catchment (Fig. 4), regolith particle size and hence, permeability, varies considerably but in a somewhat consistent manner with depth. This anisotropy has been observed similarly in regolith profiles from Malawi (Chilton & Smith-Carington, 1984; Chilton & Foster, 1995) and Nigeria (Hazell et al., 1992). Differences among the four presented profiles provide some measure of the lateral heterogeneity that also occurs in residual soils. The primary aquifer is located in the saturated, coarse-grained sections at the base of the regolith. A crude estimate of the hydraulic conductivity of this aquifer was gained using particle-size data and a relationship between hydraulic conductivity and poorly-sorted sediment demonstrated recently by Shepherd (1989). Results are given in Table 1.

3.2 *Piezometer tests*

Slug and bail tests were successfully performed at six sites in both catchments and were interpreted using a computational routine known as 'AQTESOLV' (Duffield & Rumbaugh, 1991). A summary of the results is shown on Table 1. Inflow (slug) and outflow (bail) tests gave comparable results with values for hydraulic conductivity (*K*) averaging 0.3 m day^{-1}. Results from a recent interpretation of 60 pumping tests in shallow, regolith wells (Taylor & Howard, 1995a) from Mukono District in southeastern Uganda (Fig. 1) are included for comparison. The range in *K* values, particularly in the large data set from Mukono District (Fig. 1), stems from the heterogeneity of regolith alluded to above.

Table 1. Estimates of hydraulic conductivity in the regolith aquifer in Uganda.

Method	No. (sites)	Range K m day^{-1}	Mean K m day^{-1}	Median K m day^{-1}	Location
Grain size analysis	7	0.1 to 6	2	0.9	Aroca
Slug/Bail tests	6	0.03 to 0.7	0.3	0.2 to 0.3	Aroca/Nyabisheki
Short pumping tests	60	0.005 to 10	0.3	0.08	Mukono District

In general, results from grain size analyses are an order of magnitude higher than those determined from piezometer tests. This is expected given that the former analysis specifically targets the coarsest grained material while the piezometer tests utilize the entire screened thickness of the regolith. The hydraulic conductivity of the regolith aquifer in Uganda consequently rests somewhere in the vicinity of between 0.3 and 3 m day^{-1}, a range that compares well with results obtained for regolith aquifers in Malawi, 0.05 to 1.5 m day^{-1} (Chilton &-Smith-Carington, 1984), Zimbabwe, 0.1 m day^{-1} (Houston & Lewis, 1988) and Nigeria, 1.2 m day^{-1} (Owoade, 1989).

4 HYDRAULIC RELATIONSHIP BETWEEN REGOLITH AND BEDROCK AQUIFERS

Groundwater flow in unweathered crystalline bedrock is entirely due to the intensity and extent of fracture development. In Uganda, packer testing of boreholes (Howard et al., 1992) found that fractures occur throughout the depth of the well but often exhibit a relatively low hydraulic conductivity (< 0.1 m day^{-1}) and yield very little water. The low storativity of the fractured bedrock aquifer has led workers not only in Uganda (Hydrogeology-Uganda Phase I, 1989) but also in Malawi (Chilton & Smith-Carington, 1984) and Zimbabwe (Houston & Lewis, 1988) to surmise that the regolith provides storage for boreholes drawing water from bedrock fractures. Houston (1988), however, was unable to find a correlation between regolith thickness and borehole yield to substantiate this assumption of hydraulic continuity between the two aquifers. Evidence from piezometer tests and grain size analyses of the regolith reveals that it is significantly more permeable than the underlying bedrock and that, where a substantial thickness exists, it is likely to provide a significantly better aquifer. Therefore, an important question to be addressed is whether the two aquifers behave independently, or whether the systems are sufficiently linked hydraulically to allow vertical leakage to sustain the long-term development of the weaker bedrock aquifer.

The hydraulic relationship between the regolith and bedrock aquifers was evaluated in three ways: Analysis of short-term pumping tests of bedrock wells, examination of long-term pumping tests of bedrock wells with a regolith monitoring well, and interpretation of inorganic and isotope hydrochemistry.

4.1 *Short-term pumping tests*

Results from 34 pumping tests in both catchments show that many drawdown responses deviate from the classical 'Theis' confined aquifer response (Theis, 1935) exhibiting a significant reduction in the rate of drawdown with time. Stabilization of well drawdown reflects the contribution of unconfined storage inputs from one or a combination of: 1. Microfissures from within the consolidated bedrock, 2. An overlying aquifer, or 3. A recharge boundary (e.g. surface water). The existence of the regolith aquifer combined with generally poor curve matches found using the fractured aquifer solution of Moench (1984) and the paucity of surface water bodies, suggest that the stabilised drawdown response results from vertical leakage of water from an overlying aquifer (i.e. a 'leaky' condition). Good curve matches were obtained for a number of the tests (Fig. 5) using the leaky-aquifer method of Moench (1985). This method is able to account for well bore storage effects which can be significant at early times during short-term tests, especially when aquifer transmissivity and storage are small.

Leaky aquifer conditions were observed in the majority of tests conducted in both study areas (Table 2). A sampling of pumping tests conducted in boreholes in Mukono District shows similarly the predominance of leaky aquifer conditions in bedrock wells. In the regolith and fractured bedrock aquifer system of Zimbabwe, Houston & Lewis (1988) also observed leaky aquifer behaviour in bedrock wells though the frequency of the response was less, around 40%. A comparison of bedrock transmissivities under leaky aquifer conditions with the few confined and fractured aquifer interpretations shows only small differences between median values. The data therefore suggest that while the hydraulic continuity may be sufficient to

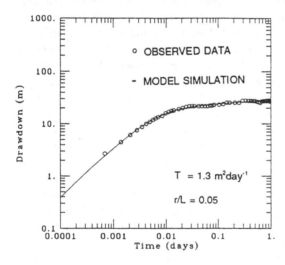

Figure 5. Pumping test interpretation of a bedrock well in the Aroca catchment (CD722) using the Moench (1984) leaky aquifer solution in AQTESOLV.

Table 2. Summary of pumping test interpretations conducted in bedrock wells.

Location	Total (sites)	Leaky (sites)	Median T m² day⁻¹	Confined (sites)	Median T m² day⁻¹	Fractured (sites)	Median T m² day⁻¹
Nyabisheki	23	17	0.7	2	0.13-1.2	4	1.2-2.1
Aroca	11	11	1.3	0	–	0	–
Mukono	7	4	2.6-4.1	3	0.99	0	–

allow water to move from the regolith to the bedrock aquifer, the low transmissivity of the bedrock aquifer will limit the transmission of water to relatively small quantities.

4.2 *Long-term pumping tests*

To confirm whether the regolith is a source of unconfined storage to bedrock wells, long-term pumping tests were performed at four sites in the Aroca catchment and at one site in the Nyabisheki where a regolith water table exists. In each test, the pumping phase lasted for 24 hours or more, and measurements of drawdown were made both in the pumped well and an observation well located in the regolith, 5 m from the pumped well. Results are summarised in Table 3.

It is significant that during each test, a lowering of the water table in the regolith was observed. Although measured drawdowns were slight, averaging just 7 cm, both grain-size analyses of the regolith and studies of the water balance in each catchment (Howard & Karundu, 1992; Taylor & Howard, 1995b), indicate that the regolith constitutes a considerably larger groundwater reservoir than the underlying bedrock. As such, leakage from the regolith induced from drawdown in the bedrock would be expected to cause only a minor drop in the regolith water table. In this light, it is significant to note that a strong correlation ($r^2 = 0.80$) exists between the magnitudes of the drawdown observed in each of the aquifers.

4.3 *Hydrochemistry*

The hydraulic relationship between regolith and bedrock aquifers was also investigated by examining hydrochemistry. Inorganic water chemistry can distinguish mixing between water bodies while their isotopic compositions can reveal their origin. Details regarding sampling protocols and analytical procedures used to acquire inorganic and stable isotope (^{2}H & ^{18}O) hydrochemistry have been published (Taylor & Howard, 1995c; Taylor & Howard, 1995b). The application of these data to the interconnection of the two aquifers is discussed below.

Inorganic chemistry. The major ion chemistry of regolith and bedrock groundwaters from both catchments is represented as a Durov plot in Figure 6. Across both catchments, groundwaters are dominated primarily by the bicarbonate anion. A small portion of samples from the regolith and bedrock aquifers in the Aroca exhibit relatively higher levels of chloride. In most samples, aqueous calcium and sodium exist in roughly equal proportions. Although water-rock interactions in both aquifer units may be expected to lead to a parallel geochemical evolution, the data, at the very

Table 3. Summary of long-term pumping test results.

Borehole	Yield (L s^{-1})	Pumped well drawdown (m)	Regolith drawdown (m)
GS 1943 (Aroca)	0.91	11.05	0.03
CD 2253 (Aroca)	0.54	23.09	0.10
CD 722 (Aroca)	0.76	27.52	0.12
CD 78 (Aroca)	0.81	4.32	0.03
CD 360 (Nyabisheki)	0.34	7.91	0.07

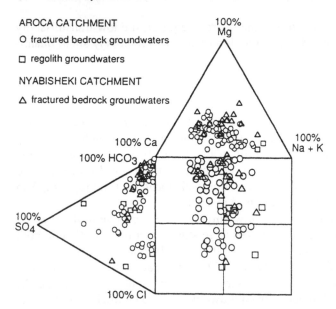

AROCA CATCHMENT
O fractured bedrock groundwaters
□ regolith groundwaters

NYABISHEKI CATCHMENT
△ fractured bedrock groundwaters

Figure 6. Durov plot of regolith and bedrock groundwaters (from Taylor & Howard, 1995c).

least, do not deny a hydraulic connection. In consideration of minor elements, Taylor & Howard (1995c) found significantly higher levels of aluminium in the regolith (1 ppm) compared to the bedrock (0.1 ppm). Assuming that higher levels of aluminium are the product of active weathering of aluminosilicate rocks in the regolith, it appears that recharge to the bedrock aquifer occurs where the aquifer units are sufficiently linked to allow water to move through the regolith and into the bedrock with minimal geochemical interaction.

Stable isotope study – Aroca catchment. The stable isotope ($^2H, ^{18}O$) composition of groundwater was determined in samples collected from 65 sites in the Aroca catchment: 54 boreholes and 11 regolith piezometers. The analytical error in each determination is $\pm 0.2‰$ for oxygen-18 and $\pm 2‰$ for deuterium. Symbols d_D and d_{18} represent the standard measurement for ^2H and ^{18}O relative to the standard, Vienna-SMOW, given by the formula:

$$d = 1000 \times (R_{SAMPLE} - R_{SMOW})/R_{SMOW} \qquad (1)$$

where R represents the ratio of heavy to light isotope (i.e. $^2H/^1H$ or $^{18}O/^{16}O$). Results are plotted in Figure 7 along with isotopic signatures from nine regional surface waters and the meteoric waterline for Entebbe, $d_D = (7.2 \pm 0.3) \cdot d_{18} + (11 \pm 6)$ (I.A.E.A., 1992).

Regolith and bedrock groundwaters plot together, over a limited range ($-7‰ < d_D < 0‰$, $-2.2‰ < d_{18} < -1.0‰$) and along the meteoric water line. Significantly, groundwaters from each aquifer possess very similar mean compositions, $d_D = -3.6‰$, $d_{18} = -1.5‰$ (regolith) and $d_D = -3.6‰$, $d_{18} = -1.7‰$ (bedrock). Since the stable isotopes of water tend to behave conservatively in low-temperature groundwater environments, the results indicate that the regolith and bedrock groundwaters are being recharged from the same source.

Groundwater recharge in Aroca catchment may result from: 1. Direct infiltration of rainfall, 2. Indirect sources such as vertical leakage from the Aroca swamp or horizontal leakage from Victoria Nile or Lake Kyoga, or 3. An earlier climatic period (i.e. palæorecharge). As shown in Figure 7, surface waters exhibit dramatically enrichment in their heavy isotope content and fall along a depressed slope (d_D/d_{18}) of 3.3 that reflects enrichment due to evaporation (Craig, 1961). This distinct separation between the composition of groundwater and surface water samples confirms that surface water bodies do not contribute significantly to recharge.

In East Africa, groundwaters derived from palæorecharge are characterised by strong isotopic depletion, indicative of recharge under historically cooler climates. The composition of palæowaters in neighbouring Sudan was found to reside in the ranges: $-70‰ < d_D < -40‰$, $-10‰ < d_{18} < -5‰$ (Darling et al., 1987). The relative enrichment in heavy isotopes noted in groundwaters from both aquifers in the Aroca catchment (Fig. 7), appears to deny the existence of significant palæorecharge. Preliminary results from the analysis of the isotopic composition of rainfall in the Aroca catchment show that the content of rainy season rainfall compares well with that of catchment groundwaters (Taylor & Howard, 1995b). Therefore, based on this tentative link between the isotopic signatures of rainy season rainfall and groundwaters, combined with the refutation of other possible sources of recharge, the stable isotope evidence strongly implies that direct infiltration of rainfall is the principal source of groundwater recharge. Since the bedrock is almost entirely concealed by the regolith in the Aroca catchment, all recharge must invariably flow through the regolith.

Tritium study – Nyabisheki catchment. Tritium (^{3}H) is the radioactive isotope of hydrogen and is incorporated in water as (^{3}H)HO. Although it is produced naturally by cosmic ray interactions with nitrogen in the atmosphere, the level of tritium in the atmosphere rose dramatically during the mid-1960s as a result of atmospheric thermonuclear testing. This influx is commonly referred to as the 'bomb pulse'. Groundwaters recharged prior to the bomb pulse exhibit background (natural) levels

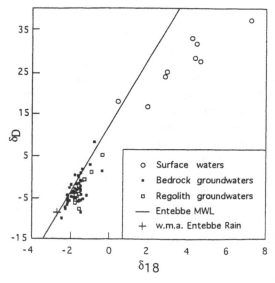

Figure 7. Isotope data plotted relative to MWL for Entebbe (from Taylor & Howard, 1995b).

of tritium while those recharged afterwards feature a tritium content significantly in excess of natural amounts. It follows that tritium can be used to constrain the date of groundwater recharge and so indicate the rate of groundwater circulation.

The average level of tritium recorded in precipitation at Entebbe prior to the bomb pulse was 11 TU (tritium units) (I.A.E.A., 1992) with a tritium unit representing 1 tritium atom in 10^{18} atoms of hydrogen. Because the half life of tritium is 12.3 years, groundwaters derived from precipitation prior to thermonuclear testing are expected to have tritium levels of less than 2 TU. In 1963, tritium levels in rainfall at Entebbe rose to 250 TU (I.A.E.A., 1992). Although a moratorium on atmospheric testing restricted tritium inputs by 1970, tritium content in precipitation observed during the 1970s was in the order of 20 TU (I.A.E.A., 1992). As a result, groundwaters recharged after 1963 are expected to have a tritium content of greater than 2 TU.

The tritium content of groundwaters from both the regolith and bedrock aquifers was determined at two sites in the southwestern part of the Nyabisheki catchment (Fig. 2) where significant aquifers (regolith saturated thickness > 5 m) were identified during drilling. For the purposes of this study, it was assumed that the level of tritium in rainfall of the Nyabisheki catchment would be comparable to that observed at Entebbe. In regolith groundwaters, the amount of tritium was 13 and 14 TU. In the underlying bedrock at the same sites, tritium was found, in both cases, to be less than 6 TU. The tritium data suggest that regolith groundwaters are modern, post 1963, while bedrock groundwaters were primarily recharged prior to 1963. This difference in the groundwater ages supports the findings from water balance studies in both catchments (Howard & Karundu, 1992; Taylor & Howard, 1995b), that the regolith represents the more active groundwater pathway.

5 RECHARGE ANALYSIS

Estimates of groundwater recharge in Uganda and other areas across equatorial Africa are severely constrained by a paucity of regional meteorological and hydrogeological records. Nevertheless, during the Phase I investigation, a soil moisture balance study was performed using a limited data set (4 year period) in the Nyabisheki catchment and suggested an average recharge rate of just 17 mm year^{-1} (Howard & Karundu, 1992). The rate, although tentative due to the limited data set upon which it was based, is comparable to rates determined in drier areas of Zimbabwe (12 mm year^{-1}) by Houston (1990). In Phase II work, a longer period was sought for soil moisture balance calculations. Furthermore, stable isotope and flow modelling studies were used to corroborate the soil moisture balance evaluation.

Groundwater recharge in the Aroca catchment, as determined by the stable isotope study described above, results from the direct infiltration of rainfall. Rainfall of 1400 mm year^{-1} in this region follows a bi-modal distribution throughout the year which stems from the movement of air masses associated with the Intertropical Convergence Zone (ITCZ). Potential evaporation in this region is in the order of 2000 mm year^{-1} and, as with other monsoonally-determined climates in equatorial Africa, restricts rainfall fed recharge by exceeding rainfall throughout most of the year (Fig. 8).

Details regarding the soil moisture balance, stable isotope and flow modelling studies employed in Phase II are described by Taylor & Howard (1995b). The annual distribution of the daily average recharge estimated from soil moisture balance calculations for the period 1988-1993 is shown in Figure 9. A daily time step was used since Howard & Lloyd (1979) have shown that longer intervals can lead to an underestimation of recharge. Significantly, the results show recharge occurs exclusively during the wet season from April to October and reflects the bi-modal nature of incoming precipitation. During the 'first' rainy season in April and May recharge is limited by large soil moisture deficits developed during a long preceding dry season. After a short 'dry' season during June and July, the 'second' rains have a considerably smaller deficit to recover before restoring field capacity in the soil zone and in-

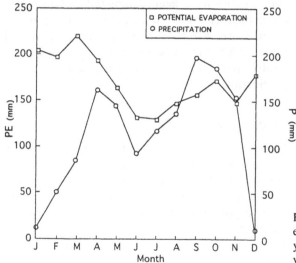

Figure 8. Distribution of potential evaporation and rainfall through the year in the Aroca catchment (monthly values).

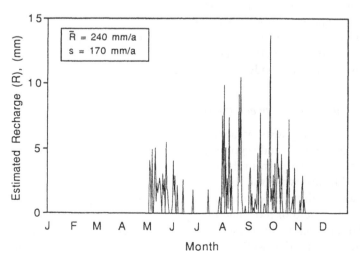

Figure 9. Distribution of estimated recharge (average daily values from 1988 to 1993) (from Taylor & Howard, 1995c).

filtrating the sub-surface as recharge. This mechanism is supported, in part, by recent stable isotope measurements of Aroca rainfall (Taylor & Howard, 1995b) as well as previous recharge evaluations across equatorial Africa which have shown that recharge appears to be restricted to intensive rainfall during the monsoons when soils are saturated and daily precipitation can exceed daily evapotranspiration (Houston, 1982; Singh et al., 1984; Houston, 1990; Adanu, 1991). In addition, Taylor & Howard (1995b) have demonstrated that the rate of recharge (200 mm year^{-1}), confirmed by flow modelling studies, exhibits a greater dependency upon the number of heavy (> 10 mm day^{-1}) rain events than the total volume of rainfall.

6   CONCLUSIONS

Hydrogeological investigations of the regolith have been completed in two catchments of Uganda which feature different bedrock lithologies, catchment topographies and climates. The overall findings of this work, therefore, tend to be representative of basement terrain in this region as a whole. Although aspects pertaining to the development, or evolution, of regolith aquifers have not been considered here, this topic will comprise the focus of future investigations.

6.1   *Hydrogeological characteristics of the regolith aquifer*

In the Aroca catchment, an area of low relief and high rainfall, the regolith aquifer was found to be regionally extensive. In contrast, lower rainfall and a more steeply sloping topography in the Nyabisheki catchment has restricted development of aquifers in the regolith. Regolith aquifers occur only in the western and southwestern portions of the catchment, which receive proportionately more rainfall and have a more gentle slope. Particle size analyses of regolith material collected during the installation of regolith piezometers show considerable heterogeneity with depth and from site to site. However, a similar progression down the profile was observed with a thin (1 to 2 m), sandy lateritic crust resting over a thicker (4 to 10 m) muddy formation which in turn overlies a coarser, muddy-sand aquifer. A combination of particle size analyses, slug tests and pumping tests of the regolith aquifer show that hydraulic conductivities typically occur between 0.3 to 3 m day^{-1}. Thus, with a saturated thickness ranging from 10 to 20 m, the regolith aquifer is often an order of magnitude more transmissive than the underlying bedrock aquifer (»1 m^2 day^{-1}).

6.2   *Degree of hydraulic interaction between the regolith and bedrock aquifer*

Evidence from short-term pumping tests of bedrock wells shows that the hydraulic interconnection between the bedrock and regolith aquifers is sufficient in 80% of the tested wells to allow vertical recharge from the regolith to wells in the bedrock. Despite this link to the regolith reservoir, the low transmissivity of wells displaying a leaky response, as with wells exhibiting confined conditions, would appear to restrict their development beyond the yield of most handpumps (10 m^3 day^{-1}). During longer (> 24 hr) pumping tests of bedrock wells, drawdown and a delayed recovery were recorded in adjacent regolith observation wells. This response confirms hydraulic connection between the regolith and underlying bedrock, a characteristic further supported by stable isotope evidence which shows clearly that both aquifers derive recharge from the same source.

6.3 *Nature of recharge to the regolith and bedrock aquifers*

Aquifer recharge varies considerably across the region. In the Aroca catchment, recharge is in the order of 200 mm year^{-1} and results from the direct infiltration of rainfall during the region's two rainy seasons. In the Nyabisheki catchment, lower rainfall and a high relief, which favours rainfall runoff, provides a significantly lower recharge rate of 17 mm year^{-1}. Analysis of tritium in groundwater samples from the Nyabisheki catchment indicates that regolith waters are modern (post 1963) while bedrock groundwaters are predominantly older (pre-1963). Flow modelling studies in the Aroca catchment (Taylor & Howard, 1995b) show that the vast majority of estimated recharge moves by way of the more permeable regolith in preference to the underlying bedrock. The studies in both catchments conclude that the regolith represents the more active groundwater pathway. As a result, wells constructed and screened exclusively in the regolith aquifer may prove to be a more viable long-term development option.

ACKNOWLEDGEMENTS

The authors gratefully acknowledge numerous colleagues who assisted in the study. Special thanks go to John Karundu, Callist Tindimugaya and Edwardmartin Rwarinda of the Hydrogeology Section, Directorate for Water Development in Uganda as well as Sean Salvatori and Kit Soo of the University of Toronto. Stable isotope and tritium samples were analysed by the Environmental Isotope Laboratory of University of Waterloo, Canada. Analysis of aqueous inorganic chemistry was performed in the Fluid-inclusion Laboratory of University of Toronto. Pumping test data from southeastern Uganda were supplied by the Water Resources Section of the Rural Water and Sanitation Project (RUWASA), sponsored by the Danish International Development Agency. The work was supported by research grants from the International Development Research Centre (IDRC, Canada) and Natural Sciences and Engineering Research Council of Canada (NSERC). During the preparation of this manuscript, the principal author benefitted in particular from the provision of a Young Canadian Researchers Award from IDRC and an Open Fellowship from the University of Toronto.

REFERENCES

Acworth, R.I. 1987. The development of crystalline basement aquifers in a tropical environment. *Quarterly Journal of Engineering Geology* (London) 20: 265-272.

Adanu, E.A. 1991. Source and recharge of groundwater in the basement terrain in the Zaria – Kaduna area, Nigeria: applying stable isotopes. *Journal of African Earth Sciences* 15(3/4): 229-234.

Chilton, P.J. & Smith-Carington, A.K. 1984. Characteristics of the weathered basement aquifer in Malawi in relation to rural water supplies. In: Challenges in African Hydrology and Water Resources. *Proc. of the Harare Symp., July 1984.* I.A.H. Series Publication 144: 15-23.

Chilton, P.J. & Foster, S.S.D. 1995. Hydrogeological characteristics and water-supply potential of basement aquifers in tropical Africa. *Hydrogeology Journal* 3(1): 3-49.

Craig, C. 1961. Isotopic variations in meteoric waters. *Science* 133: 1702-1703.

Duffield, G.M. & Rumbaugh III, J.O. 1991. *'AQTESOLV' Aquifer Test Solver*, Ver. 1.00. Geraghty & Miller Inc., Virginia, USA..

Darling, W.G., Edmunds, W.M., Kinniburgh, D.G. & Kotoub, S. 1987. Sources of recharge to the basal Nubian sandstone aquifer, Butana Region, Sudan. In: *Isotope Techniques in Water Resources Development*, IAEA-SM-299/177, IAEA, Vienna, 205-225.

Freeze, R.A. & Cherry, J. 1979. *Groundwater*. Prentice-Hall, New Jersey.

Hazell, J.R.T., Cratchley, C.R. & Jones, C.R.C. 1992. The hydrogeology of crystalline basement aquifers in northern Nigeria and geophysical techniques used in their exploration. In: *Hydrogeology of crystalline basement aquifers in Africa*. Geological Society Special Pub. 66: 155-182.

Houston, J.F.T. 1982. Rainfall and recharge to a dolomite aquifer in a semi-arid climate at Kabwe, Zambia. *Journal of Hydrology* 59: 173-187.

Houston, J.F.T. & Lewis, R.T. 1989. The Victoria Province drought relief project, II. Borehole yield relationships. *Ground Water* 26(4): 418-426.

Houston, J. 1990. Rainfall – runoff – recharge relationships in the basement rocks of Zimbabwe. In: *Groundwater Recharge; a guide to understanding and estimating natural recharge*, IAH 8: Chapter 21.

Howard, K.W.F. & Lloyd, J.W. 1979. The sensitivity of parameters in the Penman evaporation equations and direct recharge balance. *Journal of Hydrology* 41: 329-344.

Howard, K.W.F. 1991. Potential over-development of fractured bedrock aquifers in the Nyabisheki catchment of south-western Uganda. In: Aquifer Overexploitation. *Proc. of the 23rd IAH Congress, Tenerife, Spain* (April, 1991): 437-440.

Howard, K.W.F., Hughes, M., Charlesworth, D.L. & Ngobi, G. 1992. Hydrogeologic evaluation of fracture permeability in crystalline basement aquifers of Uganda. *Journal of Applied Hydrogeology* 1: 55-65.

Howard, K.W.F. & Karundu, J. 1992. Constraints on the development of basement aquifers in east Africa – water balance implications and the role of the regolith. *Journal of Hydrology* 139: 183-196.

Hydrogeology-Uganda Phase I. 1989. Hydrogeology of fractured bedrock systems in southwest Uganda. Final report to the Water Development Department, Ministry of Water and Mineral Development, Republic of Uganda (October, 1989).

Hydrogeology-Uganda Phase II. 1994. A hydrogeological and socio-economic examination of the regolith and fractured bedrock aquifer systems of the Aroca catchment in Apac District and the Nyabisheki catchment in Mbarara District. Final report to the Directorate of Water Development, Ministry of Natural Resources, Republic of Uganda (May, 1994).

International Atomic Energy Agency. 1992. Statistical treatment of data on environmental isotopes in precipitation. Technical Report Series no. 331, IAEA, Vienna.

Jones, M.J. 1985. The weathered zone aquifers of the basement complex areas of Africa. *Quarterly Journal of Engineering Geology* (London) 18: 35-46.

Leggo, P.J. 1974. A geochronological study of the basement complex of Uganda. *Journal of the Geological Society of London* 130: 263-277.

Leroux, M. 1983. *The Climate of Tropical Africa*. Champion, Paris.

McFarlane, M.J., Chilton, P.J. & Lewis, M.A. 1992. Geomorphological controls on borehole yields; a statistical study in an area of basement rocks in central Malawi. In: *Hydrogeology of crystalline basement aquifers in Africa*. Geological Society Special Pub. 66: 131-154.

Moench, A.F. 1984. Double-porosity models for a fissured groundwater reservoir with fracture skin. *Water Resources Research* 20(7): 831-846.

Moench, A.F. 1985. Transient flow to a large-diameter well in an aquifer with storative semi-confining layers. *Water Resources Research* 21(8): 1121-1131.

Ollier, C.D. 1959. A two-cycle theory of tropical pedology. *Journal of Soil Science* 10(2): 137-148.

Ollier, C.D. 1994. Exploration concepts in laterite terrains. *The AusIMM Bulletin* 3: 22-27.

Owoade, A., Hutton, L.G., Moffat, W.S. & Bako, M.D. 1989. Hydrogeology and water chemistry in the weathered crystalline rocks of southwestern Nigeria. In: Groundwater Management: Quantity and Quality. *Proc. of the Benidorm Symp. (October, 1989)* 210-214.

Nahon, D. & Tardy, Y. 1992. The ferruginous laterites. In: *Regolith exploration geochemistry in tropical and sub-tropical terrains*. Handbook of Exploration Geochemistry 4: 40-55.

Shepherd, R.G. 1989. Correlations between permeability and grain-size. *Ground Water* 27(5): 633-637.

Singh, J., Wapakala, W.W. & Chebosi, P.K. 1984. Estimating groundwater recharge based on the infiltration characteristics of layered soil. In: Challenges in African Hydrology and Water Resources. *Proc. of the Harare Symp., July 1984*. IAHS Series Publication 144: 37-45.

Taylor, R.G. & Howard, K.W.F. 1995a. Averting shallow-well contamination in Uganda. In: Sustainability of water and sanitation systems. *Proc. of the 21st WEDC Conf. (Sept. 4-5, 1995, Kampala, Uganda)* 62-65.

Taylor, R.G. & Howard, K.W.F. 1995b. Groundwater recharge in the Victoria Nile basin of East Africa: support for the soil moisture balance approach using stable isotope tracers and flow modelling. *Journal of Hydrology* (in press).

Taylor, R.G. & Howard, K.W.F. 1995c. Groundwater quality in rural Uganda: Hydrochemical considerations for the development of aquifers within the basement complex of Africa. In: McCall, J. & Nash, H. (eds), *Groundwater Quality*: 31-44.

Theis, C.V. 1935. The relation between the lowering of the piezometric surface and the rate and duration of discharge of a well using groundwater storage. *Transactions of the American Geophysical Union*. 16: 519-524.

CHAPTER 10

Minimizing land and river salinization consequences of clearing in the New South Wales Mallee

S.A. PRATHAPAR
Division of Water Resources, CSIRO, Griffith, N.S.W., Australia

R.M. WILLIAMS & J.F. PUNTHAKEY
Department of Water Resources, Parramatta, N.S.W., Australia

ABSTRACT: In the NSW Mallee, better economic returns will be obtained if the land is cleared for grazing or for wheat. It is necessary to identify areas where the environmental impact is minimal before clearing is allowed. To facilitate this identification, a modelling approach comprising two stages has been used. A 200 year period was considered as an appropriate time scale in this study. Approximately 80-87% of the land was found suitable for clearing for wheat. The primary reason for the high level of suggested clearing was that the full impact of clearing on the region's groundwater levels will not be felt during the 200 years. This is due to low levels of recharge under cleared land uses, low rainfall and deep groundwater levels. The results are based on recharge rates estimated from limited data. Furthermore, issues other than land and river salination were not considered in this study. The decision makers must consider other environmental criteria before allowing land clearing and avoid the 13-20% of land identified unsuitable for clearing.

1 INTRODUCTION

In the New South Wales portion of the Mallee Region of southeastern Australia, land is primarily uncleared and used for grazing. The economic returns from such land use are very low. Better returns are obtained when the land is cleared for grazing and/or when dry land wheat is grown in rotation with grazing.

The climate in this region is semi-arid. Potential evaporation is very much in excess of natural rainfall in the region. This has led to the development of an ecosystem which uses almost all the incident rainfall, resulting in negligible recharge to the unconfined aquifer. In such an environment, a change in land use from natural vegetation *(Eucalyptus)* to wheat cropping would cause considerable change in the movement of soil water through the unsaturated zone and increase the rate of recharge. An increase in recharge will result in the rise of groundwater levels. The groundwater in the unconfined aquifer is saline (approximately 30,000 ppm). A rise in groundwater levels will bring this saline groundwater near the soil surface and increase the potential for land salinization in low lying areas. Furthermore, an increase in groundwater levels below the river beds will increase the discharge of saline groundwater into the rivers. Therefore, it is necessary to identify areas where the environmental impact of clearing is minimal, before land clearing is allowed.

To facilitate identification of such areas, a modelling approach comprising two stages has been developed. The first stage was based on a management model (LANDMAN), which accounts for the hydrogeology and economics of proposed land use practices within the region in the medium (200 years) term. LANDMAN was used to identify approximate areas suitable for clearing for wheat in rotation. The second stage was based on a transient groundwater flow model, MODFLOW by McDonald & Harbaugh (1984), to look at the land and river salinization consequences of land use practices determined by LANDMAN in more detail.

2 MATERIALS AND METHODS

2.1 *The site*

The study area is located in south-western NSW, between latitudes 33°15' S and 34°45' S and longitudes 141° and 144° (Fig. 1). The rainfall ranges between 243 and 308 mm year^{-1}; the potential evaporation is approximately 1600 mm year^{-1}. The

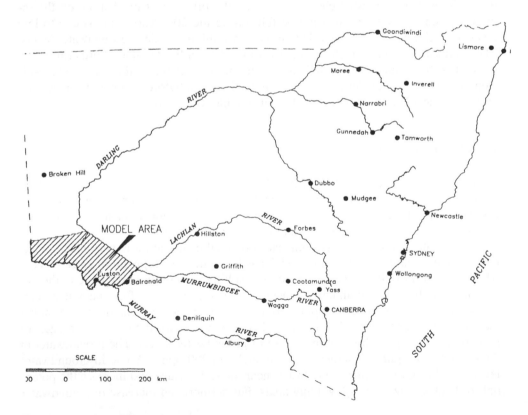

Figure 1. Location of the study area.

Murray, Murrumbidgee and Darling Rivers, which drain through the study area, receive water from rainfall in the eastern highlands of New South Wales, Queensland and Victoria.

The geology and the hydrogeology of the study area is reported by Williams et al. (1993). Five soil groups, namely, Brownish sands (BS), Clay Duplex on alluvial plains (CD), Calcareous Earth (CE), Calcareous and Siliceous Sands (CS), and Red Duplex (RD) have been identified in this area. Soils which are not associated with the above five are grouped into one category and will be referred to as Nil (NI) in this paper.

2.2 *The management model, LANDMAN*

LANDMAN was based on a two-dimensional finite difference grid of 108 rows and 63 columns. The grid is compatible with the Murray Basin Hydrogeological Model grid, defined by the Murray Darling Basin Commission. The grid axes are parallel to the directions of constant easting and northing. Each cell was of 2500 × 2500 m (ΔX and ΔY) in size. The study area was 23,281 km^2 in size and included 3724 active cells.

2.3 *Land use definitions in LANDMAN*

The land use practices considered in LANDMAN were: Uncleared grazing (U), cleared grazing (C) and cleared for wheat in rotation (W). It is feasible for a cell to have more than one soil type, and for each soil type to have all three land use practices. Therefore, the land use constraint imposed for every active cell within the region was as below:

$$Area_{ij} = \sum_p \sum_s Area_{ijsp} \tag{1}$$

where i = row number, j = column number, s = soil group (BS, CD, CS, CE, RD, NI), p = land use (U, C, W), $Area_{ij}$ = area of cell (m^2).

2.4 *Recharge definitions in LANDMAN*

The recharge rate under uncleared grazing and cleared grazing was 0.1 mm year^{-1} in this region (Cook & Walker, 1989). This is also equal to the rate at which the soil water leaves the root zone. When the land is cleared for wheat in rotation, the rate at which the soil water leaves the root zone is increased. Due to increased water content below the root zone, the matric potential distribution above the groundwater levels will be increased. It will also increase the rate at which soil water reaches the water table, until the rate at which the soil water leaves the root zone and the rate at which soil water reaches the groundwater levels are equal.

The annual recharge flux as a function of time was estimated by Cook et al. (1989) to be:

$$R_t = \int_{D\theta_0/2t}^{\infty} qf(q)dq \tag{2}$$

where, R_t = Recharge flux as function of t (mm year^{-1}), t = time since clearance (years), D = depth to groundwater (mm), θ_0 = initial volumetric water content above matric potential front, q = recharge rate (mm year^{-1}), and $f(q)$ was defined as,

$$f(q) = \frac{1}{q\sigma\sqrt{2\pi}} \exp\left[-(\ln q - \mu)^2 / 2\sigma^2\right] \tag{3}$$

where σ = log variance, μ = log mean.

μ and σ values used for the soils in the region are presented in Table 1 (Williams et al., 1993). Initial volumetric soil water content θ_0 was assumed as 0.12 for all the soil types, throughout the study area. The initial depth to the groundwater in a cell was estimated by subtracting the groundwater level (AHD) from its average ground surface elevation (AHD). The annual recharge fluxes for each soil type were summed over Y years to obtain the cumulative recharge CR_{ijs} in Y years. The CR_{ijs} was divided by $365Y$ to estimate the daily recharge flux in each soil group R_{ijsp} when the land use is W, within the cell during the Y years. For cells where the accelerated recharge front does not reach the groundwater in Y years, the recharge flux was set to 0.1 mm year^{-1}.

The recharge constraint for every active cell in LANDMAN was:

$$R_{ij} = \frac{\Sigma_s \ \Sigma_p \ Area_{ijsp} * R_{ijsp}}{\Delta X * \Delta Y} \tag{4}$$

where R = recharge flux (m day^{-1}), ΔX = horizontal cell size (m), ΔY = vertical cell size (m).

2.5   Groundwater flow definitions in LANDMAN

The unconfined aquifer underlying the study area was taken to be a single layer, comprising the Pliocene Sand and Shepparton Formation. The finite difference grid imposed on the study area was also applied to the unconfined aquifer. An explicit two-dimensional. finite difference approximation of the groundwater flow equation (Eq. 5) was used to determine the groundwater flow for every active cell in the grid:

$$h_{ij}^{n+1} = \left(1 - \frac{4T\Delta t}{S\Delta X\Delta Y}\right)h_{ij}^n + \left(\frac{4T\Delta t}{S\Delta X\Delta Y}\right) \cdot$$
$$\left(\frac{h_{i+1,j}^n + h_{i-1,j}^n + h_{i,j+1}^n + h_{i,j-1}^n}{4}\right) + \frac{R_{ij}^n \Delta t}{S} \tag{5}$$

Table 1. Means and variances of recharge flux function.

Soil group	μ	σ
Brownish sands, BS	2.62	0.76
Clay duplex on alluvial plains, CD	1.49	0.37
Calcareous earths, CE	1.83	0.37
Calcareous and siliceous sands, CS	2.93	0.95
Red duplex, RD	1.49	0.25
Nil, NI	1.49	0.25

where h^n = groundwater level at nth time step (m AHD), n = time step, T = transmissivity (m^2 day^{-1}), S = specific yield, R = recharge flux (m day^{-1}), Δt = duration of time step (day).

The hydraulic properties of the aquifer, the initial piezometric levels and the boundary conditions used in LANDMAN are presented by Prathapar et al. (1994).

2.6 *River hydraulics in LANDMAN*

The rate at which salt will be discharged into the river at the end of Y years (RSE) was estimated using the equations below:

$$RSE = 0, \quad h^n_{rij} > h^n_{ij} \tag{6}$$

$$RSE = SC * [CRIV_{ij} * (h^{n+1}_{ij} - h^n_{rij})], \quad h^n_{rij} \leq h^{n+1}_{mij} \tag{7}$$

$$CRIV_{ij} = K_r * L_{ij} * WR / TB \tag{8}$$

where $h_r{}^n$ = river head at nth time step (m AHD), SC = salt concentration in groundwater (t m^{-3}), $CRIV$ = conductance of river bed (day^{-1}), K_r = vertical hydraulic conductivity of river bed (m day^{-1}), L_{ij} = length of the river in cell (m), WR = width of the river (m), TB = thickness of river bed (m), h^{n+1}_{mij} = maximum possible head in cell ij at $n + 1$(m).

h^{n+1}_{mij} was defined as:

$$h^{n+1}_{mij} = h^n + \left[\frac{\sum_s R_{ijs} * Area_{ijs}}{S * Area_{ij}} \right] \tag{9}$$

The assumed values for SC, K_r, WR and TB were 0.03 t m^{-3} (30,000 ppm), 0.01 m day^{-1}, 100 m and 5 m respectively. The length of a river within the river cells is presented in Prathapar et al. (1994). The rate at which salt will be discharged in a particular year (RSA) was estimated by Equation (10):

$$RSA_y = \left(\frac{RSE}{Y} \right) * y \tag{10}$$

where y = year of interest (1,2,........Y).

2.7 *Economics of river salinization*

The net present value of the cost of river salinization during Y years was estimated by Equation (11):

$$RS_{ij} = \sum_y \frac{RSA_y * COSTRV_{ij}}{1.07^y} \tag{11}$$

where $COSTRV_{ij}$ = cost of salt inflow in cell ij (A\$ t^{-1} year^{-1}).

The cost of river salinization for a ton of salt discharge in a year into a river cell is presented in Prathapar et al. (1994) The value 1.07 in the denominator represents a 7% annual discount rate.

2.8 *Economics of land salinization*

In each cell the elevation (AHD) which represents 10% of the lowest topography ($DIFTOP_{ij}$), was identified. We assumed that if h_{mij}^{n+1} was within 2 m of $DIFTOP_{ij}$, then the cell will be salinised, and the cost of land salinization of the cell will be 20% of the gross return obtainable.

Thus, the cost of land salinization in a cell LS_{ij} was estimated as below:

$$LS_{ij} = GR_{ij} * 0.2 , \quad DIFTOP_{ij} - h_{mij}^{n+1} < 2 \tag{12}$$

2.9 *Estimating net returns in LANDMAN*

The net present values in 1991 dollars (NPV) determined at a discount rate of 7 % over y years were used to estimate the gross returns (GR) in a cell for the land use practices considered:

$$GR_{ij} = \Sigma_s \Sigma_p Area_{ijsp} * NPV_p \tag{13}$$

where NR = net return, (A\$), LS = land salinization cost (A\$), RS = river salinization cost (A\$).

Subsequently the net return for a cell (NR_{ij}) was determined by subtracting the sum of land and river salinization costs in the cell from the gross return.

2.10 *Objective functions in LANDMAN*

Two objective functions were used in LANDMAN. The first objective was to maximise net returns from the study area ($F1$) during the Y years, that is, maximize:

$$F1 = \Sigma_i \Sigma_j NR_{ij} \tag{14}$$

The second objective was to minimize the losses ($F2$) due to land and river salinization during the Y years, that is, minimize:

$$F2 = \Sigma_i \Sigma_j LS_{ij} + RS_{ij} \tag{15}$$

LANDMAN solved these two objectives consecutively. A 200 year period was considered as an appropriate time scale (i.e. $Y = 200$) in this study. LANDMAN was coded in a high level programming language, GAMS (Brooke et al., 1988).

LANDMAN determines land use for each cell based on the recharge rates and the net present values (NPV). The recharge rate for uncleared grazing and cleared grazing was 0.1 mm year^{-1} as determined by Cook et al. (1989). So, whether the land in a cell is cleared for grazing or uncleared for grazing, the recharge rate will be the same. The recharge rate under wheat in rotation varied with soil type and depth to the groundwater. Consequently, the environmental impact from wheat in rotation will be different in each cell. The NPV for the two grazing practices were also different, and depended on the expected commodity prices.

Therefore, for the purpose of this study, only the grazing practice with the higher NPV and wheat in rotation have to be considered in each LANDMAN run.

When expected commodity prices are 25% lower, cleared grazing and wheat in rotation result in losses. Under such circumstances, it is not necessary to run LANDMAN. Therefore the first run (Run 1) included grazing (U) and wheat in rotation (W) and used 'average' NPVs. The second run (Run 2) included cleared for grazing (C) and wheat in rotation (W) and used 'high' NPVs. The 'average' NPVs for U and W were A$18.85 and A$191.21 respectively. The high NPVs for C and W were A$63.58 and A$460.26 respectively. We recognise that the NPV at the end of 200 years will be of limited value in any study. However, for this study the relative ranking of NPV among the three land uses is the critical input, and not absolute values in dollars. In essence, we assume that wheat in rotation will result in higher returns than uncleared or cleared grazing only.

Since the specific yield was considered as the most sensitive aquifer parameter (the specific yield value for Runs 1 & 2 was 0.3), a final run (Run 3) was made for a 200 year period using a specific yield value of 0.15 and conditions for Run 2.

2.11 *The transient groundwater flow model, MODFLOW*

Since it was required to combine the economic aspects of land use and the hydrogeology of the study area, LANDMAN was constructed as a single time step management model rather than a transient groundwater simulation model. The use of a single time step for periods as long as 200 years will result in a number of errors in groundwater flow estimates. Therefore, a groundwater simulation model, MODFLOW, was used to corroborate the results determined by LANDMAN. Except for transmissivities, the aquifer parameters initial piezometric levels and the boundary conditions were the same as those used in LANDMAN. Instead of transmissivities, horizontal hydraulic conductivity values were used. MODFLOW was run with a 12.18 day time step (6000 time steps in 200 years).

Three MODFLOW runs were made in this study. Initially, it was run with the assumption that land use W was not allowed anywhere in the study area (MODFLOW-1). The recharge rate used for this run was 0.1 mm year^{-1}. The second run, MODFLOW-2, was made with varying annual recharge rates for land use practices determined from Run 2. The third run, MODFLOW-3, was made with varying annual recharge rates for land use practices determined from Run 3. The varying annual recharge rates for each cell for MODFLOW-2 and MODFLOW-3 were estimated with Equations (2) and (3).

3 RESULTS AND DISCUSSIONS

3.1 *Recharge in land cleared for wheat*

The impact on the groundwater levels from land cleared for wheat within the study area is not immediate. The initial depth to groundwater levels within the study area ranged from –10 m to 76 m. The negative depth was within the northwest corner of the study area where salt lakes are found. The 76 m depth was along the Ivanhoe ridge which runs in a NE-SW direction across the middle of the study area. The average depth to the groundwater level was approximately 22.4 m within the study area.

For soil types RD, CD and CE, the recharge rate from land use W is not expected to exceed the natural recharge rate in 200 years if the depth to groundwater is greater than 26, 36 and 54 m respectively. For other conditions, it is expected that the recharge rate from land use W will exceed the natural recharge rate within 200 years. Similar observations were made by Allison et al. (1990) for the area southwest of the study area.

The steady state (maximum) recharge rate for the soil types BS, CD, CE, CS, RD and NI were 18.3, 4.8, 6.7, 29.4, 4.6 and 4.6 mm year^{-1} respectively. For any soil type and depth to groundwater combinations, with the exception of soil type CS with depth to groundwater less than 13 m, the recharge rate will not achieve steady-state within a 200 year time frame.

3.2 Area for cleared wheat

Total area identified for the land use practices as determined by LANDMAN runs are summarised in Table 2. Since the area unsuitable for clearing for wheat in rotation includes areas targetted to minimize land salinization and river salinization, these are identified separately.

It is noted that the areas of uncleared and cleared landuse for grazing were the same. This is due to the NPV of both grazing land uses being less than that of the corresponding NPV for land use W. Thus, when minimizing cost of salinization, LANDMAN chose the same cells for grazing to minimize land and river salinization.

Based on the results from Runs 2 and 3, area suitable for land use W in each soil group is summarised in Table 3.

Table 2. Total area identified for various land use practices (ha).

Run No.	Area identified for grazing to minimize salinization losses			Area suitable for cleared wheat
	River salinization	Land salinization	Total	
1	97,490	200,625	298,115	1,925,016
2	97,490	200,625	298,115	1,925,016
3	84,359	375,625	459,984	1,867,500

Table 3. Area recommended for clearing for wheat by soil group (ha) determined by Runs 2 and 3.

Soil group	Total area	Area recommended for clearing		% of total area recommended for wheat	
		Run 2	Run 3	Run 2	Run 3
BS	247,984	237,953	228,693	96	92
CD	409,215	344,608	322,865	84	79
CE	1,140,234	1,057,874	991,799	93	87
CS	149,255	119,414	70,054	80	47
RD	158,112	131,151	125,975	83	80
NIL	222,684	138,369	128,114	62	58

With Run 2, approximately 2,029,369 ha (87%) of the total area of 2,327,500 ha were found suitable for clearing for wheat in rotation. With Run 3, approximately 1,867,500 ha (80%) were found suitable for clearing for wheat in rotation. The distribution of area found suitable for land use W with LANDMAN Run 3 is presented in Figure 2.

LANDMAN chose grazing for cells where depth to groundwater at the end of 200 years was less than 2 m, if all the cells were cleared for wheat. This is because 20% NPV of grazing was always less than 20% NPV of cleared for wheat in rotation. Nevertheless, not all the cells where depth to groundwater at the end of 200 years was less than 2 m will have groundwater within 2 m from the soil surface, because such cells were identified for grazing during minimisation of salinization costs. Thus, the cost of land salinization reported in column 3 of Table 4 reflects the reduction in returns due to grazing in such cells assuming they will be salinised. Therefore, the cost of salinization estimated by LANDMAN should be considered as the maximum land salinization costs which may incur if such cells were grazed. The maximum cost of land salinization as a percent of maximum net returns obtainable was 0.40% in Run 3.

The primary reason for LANDMAN producing a high level of clearing for wheat was that the full impact of clearing on the groundwater levels will not be felt during the 200 year period. The cumulative recharge rate in most of the areas will be less than 1000 mm in 200 years. The average recharge rate will increase by 4 to 5 fold, and the cumulative recharge volume will increase by 1 to 2 orders of magnitude (Table 5). This is due to low levels of recharge under cleared land uses, low rainfall and deep groundwater levels. The res lts are based on recharge rates estimated from limited data. Furthermore, issues otl er than land and river salination were not consid-

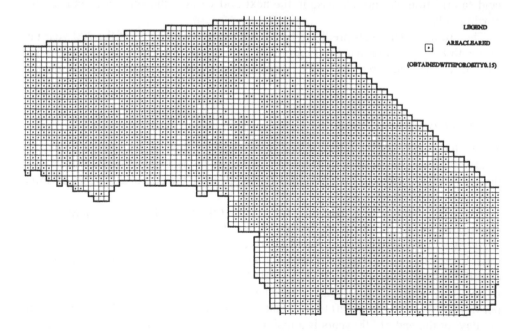

Figure 2. Land use identified for clearing for wheat with LANDMAN Run 3.

Table 4. Maximum net returns and land salinization costs (NPV A$).

Run no.	Net returns	Cost of land salinization
1	392,899,000	756,000
2	995,776,000	2,548,000
3	364,339,000	1,416,000

Table 5. Comparison of recharge rates at the end of 200 years and cumulative recharge volume in 200 years.

	Recharge rate ($m^3 day^{-1}$)	Average recharge rate (mm $year^{-1}$)	Cumulative volume ($m^3 day^{-1}$)
Grazing only	6,250	0.1	0.456-E09
Run 2 land use	31,900	0.5	0.126-E11
Run 3 land use	26,700	0.42	0.978-E10

ered in this study. The decision makers must consider other environmental criteria before allowing land clearing and avoid the 13-20% of land identified unsuitable for clearing.

To evaluate the results from LANDMAN runs, it is appropriate to consider the results from MODFLOW runs at this stage. The simulated groundwater levels after 200 years from the MODFLOW-1 run were very similar to the initial groundwater levels. This indicates that the groundwater levels within the study area are, at present, in a steady state. Therefore, it is reasonable to conclude that the area susceptible to land salinization will not increase in the next 200 years if the study area is used for grazing only.

The groundwater levels predicted by MODFLOW-3 at the end of the 200 year period with Run 3 are presented in Figure 3. A groundwater mound is developing in the NE part of the study area due to the Ivanhoe ridge posing a barrier to the groundwater flows.

3.3 *Land use and river salinization.*

The recharge rates and volumes to and from the river cells as determined by MODFLOW runs are summarised in Table 6. Clearing for wheat increased the cumulative discharge in 200 years by 60% and the discharge rate at the end of 200 years by three fold.

Assuming that land was used only for grazing (cleared and uncleared) and the discharge into the river cells took place upstream of Mildura, during the 200 years, 4.0×10^7 tons of salt would have entered the river system within the study area. In contrast, if the land use identified either with Run 2 or 3 was adopted, the total amount of salt discharged during the 200 year period would be 6.3×10^7 and 6.0×10^7 tons respectively. Further, the increase in discharge rate from 16.6 Ml d^{-1} (498 tons of salt per day) to 47.4 Ml d^{-1} (1420 tons of salt per day) or 49.3 Ml d^{-1} (1480 tons of salt per day) at the end of 200 years is a major concern. Increase in salt discharge of 500 td^{-1} will increase the river salinity at Mildura by 65 μS cm^{-1}, and an increased dis-

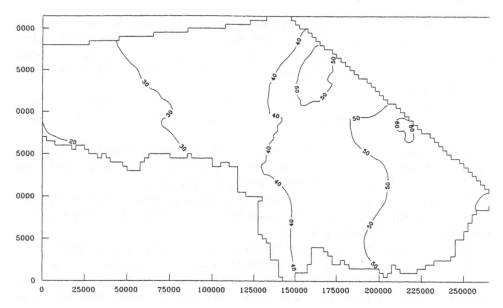

Figure 3. Groundwater levels determined by MODFLOW-3 at the end of 200 years, assuming land use determined with LANDMAN Run 3.

Table 6. Summary of recharge rates and volumes to/from the river cells.

Run no.	Recharge or discharge	Rate at the end of 200 years (m d^{-1})	Cumulative volumes in 200 years (m^3)
1	Recharge	17,400	0.17 – E10
2	Recharge	9,800	0.14 – E10
3	Recharge	8,000	0.12 – E10
1	Discharge	16,600	0.13 – E10
2	Discharge	47,400	0.21 – E10
3	Discharge	49,300	0.20 – E10

charge of 1480 td^{-1} will increase the river salinity by 194 µS cm^{-1} at the end of 200 years. It is noted that the median river salinity during 1978-1986 at Mildura was 400 µS cm^{-1} (Mackay et al., 1988).

4 CONCLUSIONS

The following conclusions could be made from this study.

– The area susceptible to land salinization will not increase in the next 200 years if the study area is used for uncleared or cleared grazing only.

– Out of the total area of 2,327,500 ha, approximately 80-87 % was found suitable for clearing for wheat. The primary reason for the high level of clearing for wheat was that the full impact of clearing on the groundwater levels will not be felt during the 200 year period. The average recharge rate will increase from 0.1 mm year^{-1} to

0.5 mm year^{-1}, and the cumulative recharge volume will increase by 1 to 2 orders of magnitude in 200 years if the recommended area is cleared for wheat in rotation. The groundwater levels will increase across the study area, and a groundwater mound will develop in the NE part of the study area due to the Ivanhoe ridge, posing a barrier to the groundwater flows.

– The maximum cost of land salinization as a percent of maximum net returns obtainable was 0.40% if the recommended level of clearing for wheat is implemented.

– If the recommended level of clearing for wheat is implemented, the total discharge of salt into the river cells during the 200 years will increase from 500 td^{-1} to 1480 td^{-1}. This will increase the river salinity by 194 μS cm^{-1}.

– The increase in cost of river salinization upstream of Mildura, for grazing only and for recommended land use at the end of 200 years, will be 7.34 and 21.8 million Australian dollars per year respectively.

Since the maximum recharge rate for BS and CS is significantly higher than the other heavier soil type, it may be appropriate to avoid clearing natural vegetation from these soil types. Further, the decision makers may want to consider environmental (e.g. fauna corridor) criteria other than land and river salinization before allowing clearing.

REFERENCES

Allison, G.B., Cook, P.G., Barnett, S.R., Walker, G.R., Jolly, I.D. & Hughes, M.W. 1990. Land clearance and river salinization in the western Murray Basin, Australia. *J. Hydrol.* 119: 1-20.

Barnett, S.R. 1989. The effect of land clearance in the Mallee region on River Murray salinity and land salinization. *BMR Journal of Australian Geology and Geophysics* 11: 205-208.

Brooke, A., Kendrick, D. & Meeraus, A. 1988. *GAMS: A Users Guide.* California: The Scientific Press.

Cook, P.G. 1989. Estimating regional groundwater recharge in the Western Murray Basin for inclusion in a groundwater model. Report No. 11, Centre for Research in Groundwater Processes, CSIRO Division of Water Resources.

Cook, P.G., Walker, G.R. & Jolly, I.D. 1989. Spatial variability of groundwater recharge in a semi arid region. *J. Hydrol.* 111: 195-212.

Mackay, N. Hillman, T. & Rolls, J. 1988. Water quality of the River Murray: Review of monitoring 1976-1988. Water Qual. Rep. No. 1., Murray Darling Basin Commission.

McDonald, M.G. & Harbaugh, A.W. 1984. A modular three–dimensional finite difference groundwater flow model. USGS, Open file report. 83-875.

Prathapar, S.A., Williams, R.M. & Punthakey, J.F. 1994. Optimising short to medium term salinization consequences of land clearing in the NSW Mallee, Australia. CSIRO DWR. Divisional Report. In Print.

Williams, R.M., Prathapar, S.A., Budd, G.R. & De Groot, J. 1993. Assessment of the economic impact of land clearing on land and water resources in southwestern New South Wales. Technical Report 92-028, Department of Water Resources, Parramatta.

Groundwater flow and solute transport models

CHAPTER 11

Modelling of mass transport in heterogeneous porous formations using indicator geostatistics

THOMAS PTAK
Institute of Geology, University of Tübingen, Tübingen, Germany

ABSTRACT: The transport of solutes in groundwater in a heterogeneous porous aquifer largely depends on the variability of aquifer parameters controlling the transport process. In general, for transport simulations in such aquifers, a detailed knowledge of these parameters and their spatial distribution is necessary, there is considerable uncertainty in the parameter values. A stochastic transport modelling approach can be employed to account for this uncertainty. A three-dimensional Monte Carlo type stochastic transport model was used for the analysis of tracer experiments, performed at the 'Horkheimer Insel' environmental research field site. The stochastic model is based on geostatistical parameters that were derived from measured hydraulic conductivity data using an indicator approach. The aquifer realizations needed for the Monte Carlo approach were generated with the conditional sequential indicator simulation technique. According to the comparison of transport simulation results with tracer test measurements, the model seems suitable for highly heterogeneous aquifer conditions such as at the 'Horkheimer Insel' field site.

1 INTRODUCTION

Modelling of solute transport in groundwater within a highly heterogeneous aquifer at small to intermediate transport distances generally requires a detailed knowledge of the aquifer parameters controlling the transport process, and their spatial variability in three dimensions. To obtain the model input parameters needed, subsurface investigation techniques are applied either at boreholes or indirectly from the surface.

At present, none of the available investigation techniques is able to provide a description of the subsurface structure and its properties at a resolution needed for a deterministic transport model in the case of a heterogeneous aquifer such as at the 'Horkheimer Insel' experimental field site, located in South Germany. Therefore, owing to the remaining parameter uncertainty, the transport modelling must be performed within a stochastic framework.

From stochastic groundwater flow and transport theory (e.g. Dagan, 1989), numerous closed-form analytical solutions or combined numerical-analytical methods based on perturbation approaches are available. In order to obtain the closed-form solutions, a small variance of hydraulic conductivity $(s^2_{\ln K})$ is assumed. Other limiting assumptions may be that the aquifer is infinite and the groundwater flow field is uniform and stationary. Furthermore, the analytical solutions generally only provide ensemble mean results.

For the analysis of tracer experiments performed at the 'Horkheimer Insel' test site, a three-dimensional Monte Carlo type numerical stochastic transport simulation technique, accounting for parameter variability and uncertainty, is applied, not having any of the above restrictions. The technique is suitable for highly heterogeneous aquifer conditions, and simulation results are obtained also for the near-source, non-ergodic stage of plume development. At this stage, an individual stochastic realization of the transport process, and consequently also the transport in a real aquifer, is likely to deviate from the ensemble mean transport behaviour. In this paper, the numerical stochastic transport simulation technique is presented and transport simulation results are compared with measurements from a non-reactive solute field tracer test conducted at a transport distance of up to about 20 m.

2   FIELD SITE AND EXPERIMENTS

2.1   *Field site*

The experimental field site is located on the 'Horkheimer Insel', north of Stuttgart, in the Neckar valley between the Neckar River and the Neckar Canal. Figure 1 shows the site map.

The aquifer is formed by a 2.5-4 m thick sequence of poorly sorted alluvial sand and gravel deposits of Holocene age. It is overlain by 5-6 m of clayey flood deposits. The saturated thickness of the generally unconfined aquifer amounts to about 3 m. From pumping tests, the geometric mean hydraulic conductivity was determined to be 0.012 m s^{-1}. The underlying Triassic clay- and limestone bedrock formation is several orders of magnitude less permeable and can be considered hydraulically tight. Groundwater flow is directed mainly to the north with a mean gradient of 0.001. The test site is presently equipped with 45 sampling and monitoring wells (Fig. 1), most of them with a diameter of 150mm, to also allow for aquifer testing. A detailed description of test site installations and experimental instrumentation is given in Hofmann et al. (1991) and in Ptak & Teutsch (1994).

During construction of the wells, a 100 mm diameter core barrel was used to obtain aquifer material samples. Using Beyer's (1964) empirical relationship between hydraulic conductivity and grain size distributions from sieve analyses of the aquifer material:

$$K = c(u) \cdot d_{10}^2 \qquad (1)$$

where $c(u)$ is an empirical constant, u is defined as d_{60}/d_{10} and d_{10} and d_{60} are the diameters of the grains where 90% and 40% respectively of the sample mass are retained, the variance $s^2_{\ln K}$ of the log-transformed hydraulic conductivity value K was estimated to be 2.34 (Schad, 1993). This is characteristic for a heterogeneous aquifer.

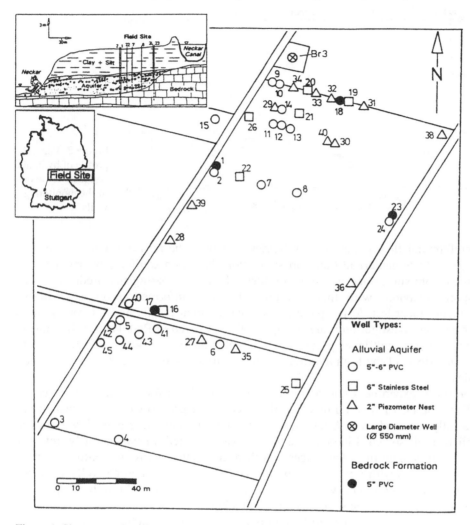

Figure 1. Site map and well location.

As an example, Figure 2 shows the vertical profile of hydraulic conductivity obtained from grain size distributions at one of the monitoring wells. It follows from the distinct variability of the hydraulic conductivity, observed in the vertical direction, that a three-dimensional transport modelling approach is needed, since at the short transport distances (of up to about 20 m) investigated, it is likely that this variability will cause a significantly irregular spread of a solute plume within the aquifer.

2.2 *Tracer tests*

Forced as well as natural gradient tracer tests have been conducted at the test site. First, the tracer tests can serve as an investigation method, yielding information on aquifer transport parameters and on aquifer structural characteristics like stratificati-

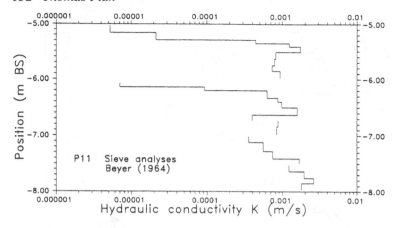

Figure 2. Vertical profile of hydraulic conductivity at one of the monitoring wells.

on, preferential flow paths or flow barriers. Second, the resulting data sets can be used to evaluate numerical transport simulations based on aquifer property data sets obtained from subsurface investigation methods, by comparing the predicted solute transport behaviour with that measured. Detailed information on experimental equipment, tracer test design parameters, measurement results and interpretation of tracer tests at investigation scales of up to 60 m is given in Ptak (1993). In some of the experiments multilevel groundwater samples have also been collected to delineate the effects of the high hydraulic conductivity contrasts observed along vertical profiles.

The tracer experiment considered here is a multi-level forced gradient tracer test (FGTT). The FGTT method involved the injection of groundwater at a constant rate into a fully screened well. After a quasi-stationary, radially divergent flow field was established, dissolved Fluresceine tracer was injected instantaneously into the groundwater supply line. The tracer was then distributed across the entire length of the water column within the well using a mixing pump. Surrounding wells, equipped with multi-level samplers, were used to monitor tracer breakthrough at up to ten sampling inlets across the saturated aquifer thickness.

Fluresceine was assumed to be representative for non-reactive solutes, which is in most cases admissible from a practical point of view (Käss, 1992). Fluresceine is generally reported to be among the least sorptive of all tracer dyes when the organic carbon content of the aquifer material is low. The organic carbon content of the 'Horkheimer Insel' aquifer material was estimated to be as low as 0.2 mg/g (Terton, 1994).

At the test site, the FGTT method was applied for short transport distance investigations at a scale of up to about 20 m. The design data of the FGTT experiment are summarized in Table 1.

Figure 3 shows a typical set of multilevel breakthrough curves obtained at one of the monitoring wells during the tracer test. As expected, the breakthrough curves are very different with respect to peak concentration, peak arrival time and spread, even at neighbouring sampling levels.

Owing to the variability of the measured breakthrough curves, it does not seem promising to apply an inverse model for the numerical simulation of the tracer test. Therefore, to account for the irregular tracer spread observed and for the remaining

Table 1. Design data of the forced gradient tracer test.

Tracer used	Fluoresceine
Tracer mass injected [g]	20
Tracer injection and groundwater infiltration well	P14
Groundwater infiltration rate [l s^{-1}]	4.05
Wells sampled / Number of sampling inlets with observed tracer breakthrough	P10/8, P11/9, P20/9, P21/9
Vertical distance of sampling inlets [m]	~ 0.3
Observation scale [m]	8.9-17.1

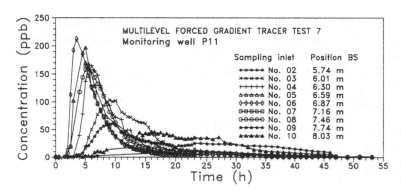

Figure 3. Multilevel tracer breakthrough curves at one of the monitoring wells.

uncertainty in aquifer parameters between monitoring wells after application of subsurface investigation methods, a Monte Carlo type numerical stochastic transport simulation technique was applied as described below.

3 THREE-DIMENSIONAL NUMERICAL STOCHASTIC TRANSPORT SIMULATION TECHNIQUE

The three-dimensional Monte Carlo type transport simulation technique comprises three steps. First, as it is generally agreed that non-reactive solute transport in heterogeneous porous formations is dominated by the spatial structure of the hydraulic conductivity field, a geostatistical analysis of hydraulic conductivity was performed. Here, hydraulic conductivity values obtained with Beyer's (1964) empirical method were used. Second, equiprobable three-dimensional realizations of the hydraulic conductivity field were generated using a geostatistical simulation approach. Third, flow and transport simulations were performed within each generated hydraulic conductivity field, with initial and boundary conditions and geometry according to the tracer experiment at the field site. This yielded an ensemble of realizations of the transport process which were compared with measurements from the field experiment.

3.1 *Geostatistical analysis of the hydraulic conductivity field*

Classical geostatistical approaches, applied for the analysis of the spatial structure of aquifers, are based on the assumption of a log-normal hydraulic conductivity distribution. They are consequently not necessarily sensitive for extreme values (Journel & Alabert, 1989). However, non-reactive transport in heterogeneous aquifers at short to intermediate transport distances is mainly determined by highly conductive zones forming preferential flow paths. Therefore, the geostatistical approach has to focus on the analysis of extreme values.

An alternative to the classical geostatistical methods is the indicator approach (Journel, 1983). It is able to honour extreme values and allows for consideration of more than one spatial structure as well as for non-normality of the log-transformed hydraulic conductivity data. A parametric distribution function of the data is not required.

In the indicator approach, the experimental histogram was calculated for the data set first. Then the hydraulic conductivity distribution was discretized into several sections by defining indicator threshold values (Fig. 4).

Three indicators were chosen here (Fig. 4), dividing the histogram into four sections. The indicator values z_{ci} and the corresponding values of the cumulative distribution function $F(z_{ci})$ were:

$$z_{c1} = 1.62 \times 10^{-4} \text{ m s}^{-1} \text{ (ind1)}   F(z_{c1}) = 0.12$$
$$z_{c2} = 6.82 \times 10^{-4} \text{ m s}^{-1} \text{ (ind2)}   F(z_{c2}) = 0.42$$
$$z_{c3} = 1.50 \times 10^{-3} \text{ m s}^{-1} \text{ (ind3)}   F(z_{c3}) = 0.70$$

Indicators were positioned in low as well as in high conductivity portions of the histogram, to take the structural properties both of the less permeable flow barriers and of the highly conductive preferential flow paths into account.

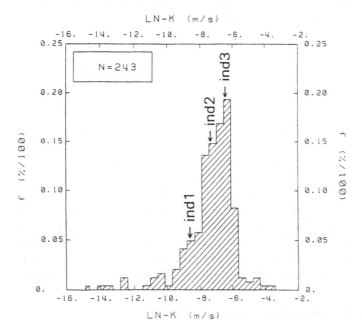

SIEB4 - Sieve Analyses (BEYER)

Figure 4. Histogram of log-transformed hydraulic conductivity values and indicators (Schad, 1993).

Subsequently, an indicator transformation of the hydraulic conductivity data was performed, yielding a binary data set for each indicator value:

$$i(\underline{x}_i, z_{ci}) = \begin{cases} 0 & \text{if } K(\underline{x}_i) > z_{ci} \\ 1 & \text{if } K(\underline{x}_i) \leq z_{ci} \end{cases} \tag{2}$$

where $i(\underline{x}_i, z_{ci})$ is the indicator transform of the hydraulic conductivity value $K(\underline{x}_i)$ at the location $\underline{x}_i$ for the indicator z_{ci}.

Then, the spatial correlation of the hydraulic conductivity field was investigated by computing experimental variograms for each of the indicators in horizontal and vertical directions.

The experimental variogram is defined by (e.g. Johnson & Dreiss, 1989):

$$\gamma_i^*(\underline{h}, z_{ci}) = \frac{1}{2N(\underline{h})} \sum_{i=1}^{N(\underline{h})} [i(\underline{x}_i + \underline{h}, z_{ci}) - i(\underline{x}_i, z_{ci})]^2 \tag{3}$$

$N(\underline{h})$ is the number of pairs of indicator transformed hydraulic conductivity values, separated by the distance $\underline{h}$.

The structural parameters of the hydraulic conductivity field were then obtained from fitting, for example, exponential variogram models:

$$\gamma(h) = c_o + \omega \cdot \left(1 - e^{-\frac{h}{\lambda}} \right) \tag{4}$$

where c_o is the nugget, ω the sill and l is the correlation length.

The detailed indicator based geostatistical evaluation is given in Schad (1993). Using 243 K-values, Schad (1993) found for the indicator variables correlation lengths between 0.1 m and 0.16 m in vertical direction, and between 1 m and 12 m for the two principal horizontal directions. The low conductivity structural elements exhibit longer correlation lengths as compared with the high conductivity zones. It is obvious, that using a classical $\ln K$-based parametric approach for the heterogeneous aquifer investigated, the structural length differences, which are assumed to exist between the sedimentary subsets defined by the indicators, were averaged out.

In the next step, the experimental histogram and the indicator variogram models are used to generate equiprobable three-dimensional spatially continuous realizations of the hydraulic conductivity field, needed for numerical flow and transport simulations.

3.2 *Generation of three-dimensional hydraulic conductivity fields*

For the non-parametric approach used here, the three-dimensional conditional sequential indicator simulation method (SIS) (Gomez-Hernandez & Srivastava, 1991) was applied. This simulation method is based on an indicator-kriging approach. It honours the values measured at monitoring well locations. Each generated realization reproduces the experimental histogram and the indicator variograms.

For the Monte Carlo transport simulations, one hundred realizations of the hydraulic conductivity field were generated. A stationarity of statistical and structural parameters was assumed. The domain simulated with the SIS-method measured 40 ×

40 × 2.4 m, extending beyond the scale of the tracer experiment. As an example, Figure 5 shows a portion of a layer from a generated hydraulic conductivity field. The structural anisotropy rendered by the SIS-method is clearly visible.

For the subsequent computation of the flow field, the domain simulated with the SIS method was extended to a total size of 2086 × 2086 × 2.4 m, using an uniform effective *K* value.

3.3 *Flow and transport simulations*

According to the conditions within the field tracer test, a stationary flow field was computed for each stochastic aquifer realization first, using the three-dimensional finite-difference flow model MODFLOW (McDonald & Harbaugh, 1984). Mass transport was then calculated employing the particle tracking program MODPATH (Pollock, 1989).

A very fine model grid resolution was chosen for the flow and transport simulations. Due to the fine resolution, the dispersivity values needed to solve the transport equation numerically were in the order of local dispersivity values. The resulting analogue for the Peclet number was defined as horizontal correlation length of the hydraulic conductivity field divided by local dispersivity. This value was large, indicating that spreading of the solute mass was mainly controlled by advection (e.g. Dagan, 1989). Therefore, following other authors (e.g. Dagan, 1990; Desbarats, 1990), effects of local dispersion were not taken into account, and concentration values were obtained from particle distributions calculated with MODPATH, which simulates advective transport.

A variable effective porosity, following a correlation function of porosity (estimated from permeameter measurements) with hydraulic conductivity, was used for the transport calculations. The correlation of porosity and hydraulic conductivity is shown in Figure 6, together with a linear regression curve and the standard devia-

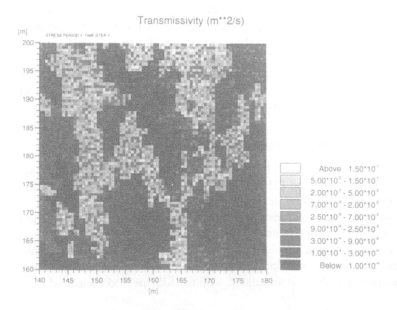

Figure 5. Portion of a model layer from a generated hydraulic conductivity field.

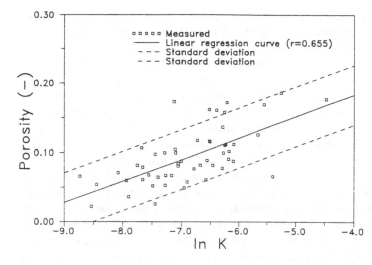

Figure 6. Correlation of porosity from permeameter measurements and hydraulic conductivity.

tion curves. Within the numerical simulations, a normally distributed random component was added to the porosity values estimated from the regression curve in order to obtain the same standard deviation of simulated porosity values as observed for the measured values. A flux weighted particle input was used for the simulation of tracer injection at the well.

It should be emphasized that no model calibration was performed. As a result, the transport simulations within each aquifer realization yielded a stochastic ensemble of particle distributions in space and time that was evaluated in terms of concentration values.

This Monte Carlo transport simulation approach has already been successfully applied in a first version in Teutsch et al. (1991) for the computation of multi-level particle breakthrough curves and in an extended version in Ptak (1993) for investigations on macrodispersivity and for the simulation of tracer experiments.

4 COMPARISON OF SIMULATED AND MEASURED RESULTS

Due to the limited number of concentration measurement locations, a computation of concentration isolines or spatial moments could not be performed. This is typical for most practical problems. Therefore, the comparison of simulated and measured tracer transport was performed using breakthrough curves.

This model testing approach is based on a comparison of effective transport parameters obtained from breakthrough curves, either directly (peak concentration, peak concentration arrival time, etc.), or from the calculation of temporal moments or from fitting analytical solutions to the breakthrough curves (mean transport velocity, longitudinal dispersivity, etc.). Each measured and numerically simulated breakthrough curve is evaluated individually. Then, effective transport parameters derived from one measured breakthrough curve are compared with a stochastic ensemble of breakthrough curve parameters, numerically simulated at a location within the model domain corresponding to the measurement position within the field tracer test.

The variance of a breakthrough curve transport parameter within the ensemble is a measure of model input parameter uncertainty. Aquifer heterogeneity is evident from spatial variability of the transport parameters.

In Figure 7, measured and simulated peak concentration arrival times (peak times) are compared at one of the monitoring wells. It is seen from Figure 7 that within the vertical profile most of the measured peak time values are close to the arithmetic ensemble mean from the stochastic simulation, and mostly within one standard deviation range of the stochastic ensemble. No systematic deviation of the measured peak times from the ensemble mean (bias) is recognized. Due to the aquifer heterogeneity, the measured as well as the ensemble mean values are variable with depth.

Because the breakthrough curve measurements were performed near a source, non-ergodic stage of tracer plume development, where the plume travelled a distance of only a few hydraulic conductivity correlation lengths, a deviation of measured breakthrough curve transport parameters from their corresponding stochastic arithmetic ensemble means, as seen in Figure 7 for example, can be expected (e.g. Graham & McLaughlin, 1989). Therefore, it is necessary to know the whole distribution of a transport parameter from numerically simulated breakthrough curves to be sure that the corresponding transport parameter from the measured breakthrough curve is within the range of the numerically simulated values.

Figure 8 gives a representative comparison of longitudinal macrodispersivities, computed using first and second temporal moments of the simulated breakthrough curves, with macrodispersivities derived from field data.

The experimental longitudinal macrodispersivies and the ensemble means of the numerically simulated values are generally close together. The experimental macrodispersivities are within the range of the numerically simulated values. They tend to be slightly higher than the numerically simulated ensemble means. This is likely to be caused by experimental effects, for example by an additional plume spread due to the time necessary for tracer injection.

Finally, Figure 9 gives a representative example of the comparison of measured and simulated peak concentration values. Again, the measured peak concentration

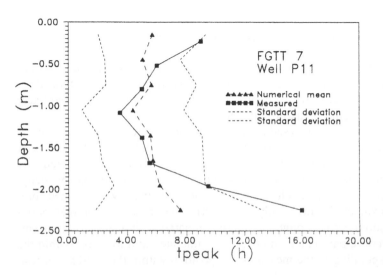

Figure 7. Example of comparison of measured and simulated peak concentration arrival times; vertical profile at one of the monitoring wells.

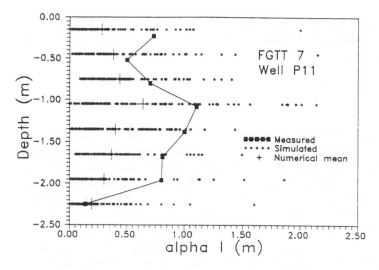

Figure 8. Example of comparison of measured and simulated longitudinal macrodispersivities; vertical profile at one of the monitoring wells.

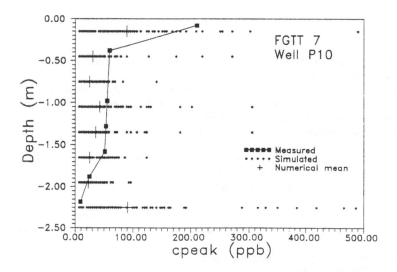

Figure 9. Example of comparison of measured and simulated breakthrough curve peak concentrations; vertical profile at one of the monitoring wells.

values at the monitoring wells are close to the stochastic arithmetic ensemble means and well within the range of the numerically simulated values.

Results of a comparable quality can also be obtained at the other monitoring wells. It follows from comparison of the effective transport parameters, obtained from breakthrough curves, that the Monte Carlo type numerical stochastic transport model yields a reliable simulation of tracer transport observed in the multi-level tracer experiment, even though the numerical model was not calibrated. Where the deviations of experimental parameter values from the numerically simulated values are considered too large, additional conditioning data values for the sequential indicator simulation method could improve the accuracy of model predictions.

5 CONCLUSIONS

A numerical Monte Carlo type stochastic transport simulation model, based on indicator geostatistical parameters, was used for a three-dimensional simulation of near-source tracer transport within a multi-level forced gradient tracer experiment. The stochastic simulation technique presented does not suffer from the restrictions of closed-form first-order solutions of stochastic transport equations. Therefore, it offers a broad field of applications, for example for the planning of remediation activities. Even for a large $s^2_{\ln K}$ value, the numerical simulation results compare well with data from the field tracer test. The simulations will be repeated using hydraulic conductivity data sets from other investigation methods to allow for an evaluation of each investigation method. Further simulations will also consider reactive transport, and more complex initial and boundary conditions as well as non-stationary flow fields.

REFERENCES

Beyer, W. 1964. Zur Beschreibung der Wasserdurchlässigkeit von Kiesen und Sanden. *Zeitschr. f. Wasserwirtschaft-Wassertechnik* 14: 165-168. Berlin.

Dagan, G. 1989. *Flow and transport in porous formations*, 465 pp. Berlin, F.R.G.: Springer Verlag.

Dagan, G. 1990. Transport in heterogeneous porous formations: Spatial moments, ergodicity, and effective dispersion. *Water Resour. Res.* 26(6): 1281-1290.

Desbarats, A.J. 1990. Macrodispersion in sand-shale sequences. *Water Resour. Res.* 26(1): 153-163.

Gomez-Hernandez, J.J. & Srivastava, R.M. 1990. ISIM3D: An ANSI-C three-dimensional multiple indicator conditional simulation program. *Comput. & Geosci.* 16(4): 395-440.

Graham, W. & McLaughlin, D. 1989. Stochastic analysis of nonstationary subsurface solute transport - 1. Unconditional moments. *Water Resour. Res.* 25(2): 215-232.

Hofmann, B., Kobus, H., Ptak, T., Schad, H. & Teutsch, G. 1991. Schadstofftransport im Untergrund, Erkundungs- und Überwachungsmethoden. PWAB, Kernforschungszentrum Karlsruhe, F.R.G.

Johnson, N.M. & Dreiss, S. 1989. Hydrostratigraphic interpretation using indicator geostatistics. *Water Resour. Res.* 25(12): 2501-2510.

Journel, A. 1983. Nonparametric estimation of spatial distributions. *Math. Geol.* 15(3): 445-463.

Journel, A. & Alabert, F.G. 1989. Non-Gaussian data expansion in Earth Sciences. *Terra Nova* 1: 123-134.

Käss, W. 1992. *Geohydrologische Markierungstechnik.* Lehrbuch der Hydrogeologie, Band 9, Verlag Gebrüder Borntraeger, Stuttgart, F.R.G.

McDonald, M.G. & Harbaugh, A.W. 1984. A modular three-dimensional finite-difference groundwater flow model. US Geological Survey Open-File Report 83-875, National Center Reston, Virginia, USA.

Pollock, D.W. 1989. Documentation of computer programs to compute and display pathlines using results from the US Geological survey modular three-dimensional finite- difference groundwater flow model. Department of the Interior, US Geological Survey, Open File Report 89-381, Reston, Virginia.

Ptak, T. 1993. Stofftransport in heterogenen Aquiferen: Felduntersuchungen und stochastische Modellierung. Dissertation, Mitteilungen des Institut für Wasserbau, Heft 80, Universität Stuttgart, F.R.G.

Ptak, T. & Teutsch, G. 1994. Forced and natural gradient tracer tests in a highly heterogeneous porous aquifer: Instrumentation and measurements. *Journal of Hydrology* 159: 79-104.

Schad, H. 1993. Geostatistical analysis of hydraulic conductivity related data based on core samples from a heterogeneous fluvial aquifer. *International Workshop on Statistics of Spatial Processes, Bari*, Italy, 27.-30. Sept. 1993.

Terton, H. 1994. *Auswirkungen der lithologischen Zusammensetzung des Horkheimer Aquifermaterials auf die Sorption von Phenantren.* Diplomarbeit, Geologisches Institut, Universität Tübingen, F.R.G.

Teutsch, G., Hofmann, B. & Ptak, T. 1991. Non-parametric stochastic simulation of groundwater transport processes in highly heterogeneous formations. *Proc. Int. Conference and Workshop on Transport and Mass Exchange Processes in Sand and Gravel Aquifers, Oct. 1-4, 1990, Ottawa, Canada*, AECL-10308 1: 224-241.

CHAPTER 12

Modelling groundwater and surface water interaction for decision support systems

IJSBRAND G. HAAGSMA & REMCO D. JOHANNS
Delft University of Technology, Department of Civil Engineering, Delft, Netherlands

ABSTRACT: This paper describes a way to study the interaction between unsteady groundwater and surface water flow. Current approaches describe the interaction at the interface in a quasi-steady way, where first the surface water movement is calculated and then the groundwater flow using the results from the surface water model. A tightly coupled model, which would mean one source code for the calculation of both the surface water part and the groundwater part is not desirable. Such a tightly coupled model would be very large and highly complex, hence difficult to maintain and not very flexible. Therefore integration of models is proposed by running existing models simultaneously, i.e. a groundwater model and a surface water model, with continuous communication. Boundary conditions are exchanged by a proposed communication interface. Time and spatial scales of the hydrological processes are important issues when models are integrated. The study of groundwater and surface water interaction is part of an approach to develop a decision support system for water management. The development of this decision support system will be discussed in this context.

1 INTRODUCTION

1.1 *Unsteady simulation*

Detailed simulation of the interaction of groundwater and surface water is seldom done and hence seldom discussed in the literature. The reason is that often one is interested in the effects of the behaviour of the surface water flow on the groundwater flow, relatively far away from the surface water. Consequently the interaction can be simulated using a relatively simple interface. Often this means that first a simulation of the surface water movement is done disregarding the presence of the groundwater, or by using average groundwater levels. After that first simulation a groundwater model is used to complete the study. The results of the surface water model are used as boundary conditions for the groundwater model. Although the calculations for both the groundwater and the surface water flows are unsteady, the interaction is assumed to be steady. In most cases this approach will prove to be accurate enough. If one is interested in the behaviour near the interface, however, a more accurate approach is needed; for example, when the water quality or quantity of extraction wells near a river reach is studied.

1.2 *Steady simulation*

Another type of simulation, that is not covered in this paper, is steady-state simulation. For example, the drainage of a polder requires modelling that accounts for the interaction between groundwater and surface water. The polder is often a drained lake or a piece of land reclaimed from the sea. The water level in the canals surrounding the polder is higher than the groundwater level in the polder, hence water leaks from these canals through the bottom into the polder causing an increase of the groundwater level in the polder. The polders are mainly used for agricultural purposes and farmers like to have constant groundwater levels for a maximum crop yield. It is not difficult to develop a model that calculates the amount of water that leaks into the polder and thus has to be discharged into the canals, again to prevent an increase of the groundwater level. The reason that this type of interaction can be modelled using a tightly coupled model, is that all the time derivatives are eliminated due to the steadiness of the problem. The absence of different time scales means that steady-state interaction can be modelled, without much difficulty, using one source code that is not very complex.

2 INTEGRATED MODELS

In the introduction tightly coupled models and loosely coupled models were distinguished. In this section we will give definitions of integrated and tightly coupled models that we will use throughout this paper. These definitions are the extremes for the wide array of possibilities in coupling models. Intermediate types of coupled models combine features of both loosely and tightly coupled models.

2.1 *Tightly coupled models*

Tightly coupled models are models that simulate different hydrological processes, such as groundwater flow and surface water flow, or water quality and water quantity. All these hydrological processes are incorporated into one source code. It is not easy to divide the source code into different sets of sub-routines in order to separate the individual processes. For small models the advantage of this approach is that one does not need to spend much time on the architecture of the software. A number of these models are developed by local water authorities for specific problems. Few large, tightly coupled models have been developed; however, there is the advantage of saving computing time, especially when a lot of the unnecessary communication present in loosely coupled models is deleted. The main disadvantage of tightly coupled models, however, is the inflexibility of these models; they are not easy to change and are not easily applied to problems other than those for which they were originally developed.

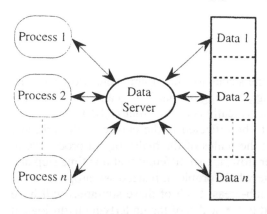

Figure 1. Inter-process communication with a data server and the optional use of a common data format.

2.2 *Loosely coupled models*

Models used for the development of the decision support system described later are loosely coupled. This means that each of the hydrological processes is described by a set of computer procedures and that communication is facilitated by a data server. The data server (Fig. 1) keeps track of all data needed and produced by all processes (Haagsma, 1995). The data server acts as an intermediary between the processes and the data files. Implementation of a data server for data that is not shared by more than one process is easy. The existing data management system can be used for that. For shared data it is recommended to use a common data format. The data server consists then of a set library routines that stores and retrieves data from files using the common data format. The data server can be very basic; however, there is no reason to make it more ingenious than necessary. Communication via a data server makes coupling of models a very flexible procedure. The authors prefer to use loosely coupled models in this way, so as to be flexible in the development of a decision support system, because development is an evolutionary process. The main effort in adding models to such a system is building the communication interface with the data server and determining what data needs to be shared.

3 MODELLING THE INTERFACE

The time scales of ground- and surface water flow differs by a few orders of magnitude. Ngo (1994) calculated that for the Netherlands the flow velocity for groundwater is in the order of 10^{-6} ms^{-1}, whereas the flow velocity in a river will typically be in the order of 1 ms^{-1}. Yen & Riggins (1991) derived a relation from which the maximum computational time scale ratio between groundwater and surface water flow can be computed. The order of this ratio is dependent on the flow depth of the surface water H, the piezometric head h, the soil porosity, which is related to the specific storage S_s and the aquifer thickness y. Using infiltration as a common boundary condition and disregarding the unsaturated zone their derivation can be formulated as:

$$O\left(\frac{T_g}{T_s}\right) \le O\left(\frac{\overline{S_s \, \overline{hy}}}{\overline{H}}\right) \tag{1}$$

where T_g and T_s are the characteristic time scales for the groundwater flow and surface water flow respectively. Yen & Riggins suggest that T_g/T_s can vary under extreme conditions from $O(1)$ to $O(10^4)$ but typically is of the order $O(10^2)$.

The huge difference in scales needs to be reflected in the choice of the grid and the time step. The reason for this is that the scales of the hydrological processes, in this paper groundwater and surface water flow, are so different that one time step and one grid size cannot be used to solve the whole problem. Instead we propose to decompose the area of interest in several sub-areas. Each of these sub-areas will have its own spatial grid and time step, related to the scales of the underlying hydrological processes.

Probably the need to use different time steps and grid sizes and the decomposition into different sub-areas is best demonstrated by an example. Consider therefore a river flowing downstream, with an abstraction well near the river as shown in Figure 2. Assuming there is a pollution spill in the upstream reach and we would like to know what influence that spill has on the quality of water abstracted from the well. The answer to that question will depend on several parameters of the hydrological system, including conductivity of both the riverbed and the aquifer, pumping rate at the well, adsorption of the pollutant, and water velocity of the water in the river. It is

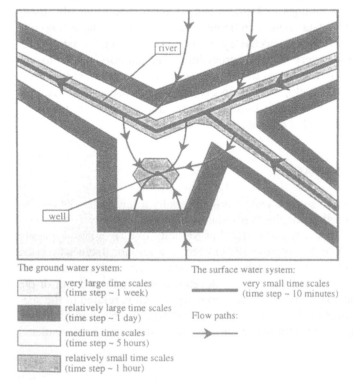

The ground water system:

- very large time scales (time step ~ 1 week)
- relatively large time scales (time step ~ 1 day)
- medium time scales (time step ~ 5 hours)
- relatively small time scales (time step ~ 1 hour)

The surface water system:

- very small time scales (time step ~ 10 minutes)

Flow paths:

→

Figure 2. Decomposition of an area into different sub-domains, each with a characteristic time scale and related time step.

clear from the picture that the boundary layer where the interaction between groundwater and surface water takes place is important. In general it is not feasible, due to constraints on computing time, to study the groundwater flow with the same time step as is used for the surface water flow. Therefore it is necessary to define a gradually increasing time step, running from small near the river reach to large in the area unaffected by the presence of the river. This procedure is similar to choosing a finer grid near a boundary layer when studying the heat transfer equation. Linking the different sub-domains involves similar considerations concerning averaging in time and space.

Two models were chosen to be integrated and serve as an example of how the interaction of groundwater and surface water flow can be modelled through integration of existing models. The main factor considered in model selection was the current use of simulation models by water authorities in the Netherlands. Geldof (1993) showed that a majority of the authorities use MODFLOW for their unsteady groundwater modelling and DUFLOW as an open channel model, so these were selected.

3.1 *Groundwater*

MODFLOW is a modular three-dimensional finite difference groundwater flow model developed by the US Geological Survey (McDonald & Harbaugh, 1983). The model simulates flow based on Darcy's Law. Its modular structure consists of a main program and a series of highly independent modules, each dealing with a specific feature of that hydrologic system. One of these modules was used to study the interaction between groundwater and surface water flow. The model provides options of a strongly implicit procedure or a slice-successive over-relaxation method to solve the discretized equations.

The partial differential equation that is used to solve the movement of groundwater of constant density through porous material, with hydraulic conductivity $K = (K_x, K_y, K_z)$ and storage coefficient S, is based on Darcy's law. Darcy's law describes a balance of spatial and temporal variation of the potentiometric head h. The equation used in MODFLOW reads:

$$\nabla \cdot (K \cdot \nabla h) - W = S \frac{\partial h}{\partial t} \tag{2}$$

The term W represents sources and/or sinks of water. W may be a function of all three space coordinates as well as time. This is the term that we will use to determine the leak term between groundwater and surface water. Figure 3 gives the conceptual representation of leakage through a riverbed into a cell. In discretised form, the flow q to cell (i, j, k) due to the presence of a river stream, with a water level H, is given by:

$$q_{i,j,k} = C_{i,j,k} (H_{i,j,k} - h_{i,j,k}) \tag{3}$$

C is the streambed conductance used to represent the stream-aquifer interconnection. It depends on the hydraulic conductivity, K_{riv}, and the thickness M of the riverbed. It further depends on the area Ar, indicating the part of the cell covered by the river

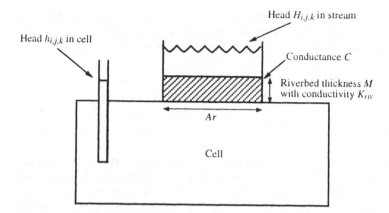

Figure 3. Conceptual representation of leakage trough a riverbed into a cell.

reach. It is the dependency on Ar that converts C into a computational quantity. It can be calculated in the following way:

$$C = \frac{K_{riv}\,Ar}{M} \qquad (4)$$

The area Ar can be at most the area covered by the whole cell, but does not include the area used for bank storage and is assumed fixed, only dependent on the flow width of the stream.

In MODFLOW the river water levels have to be known before starting the calculation, which means that there is initially only a one-way interaction. However, we will show that with a slight modification it is possible to simulate two-way interaction between groundwater and surface water.

3.2 Surface water

The surface water model DUFLOW is based on the one-dimensional equations that describe unsteady flow in open channels as derived in Abbott (1979). These equations are the mathematical translations of the laws of conservation of mass and momentum. They assume a one-dimensional flow in a straight channel. The cross-section of the channel can vary linearly between two nodes. Conservation of mass is the balance between the mass accumulating, described by the water level H in the channel and the cross-sectional storage width B, and the net discharge Q. This first balance equation can be written as:

$$B\frac{\partial H}{\partial t} + \frac{\partial Q}{\partial x} = 0 \qquad (5)$$

The second Equation (6) expresses the conservation of momentum in response to interior and exterior forces like friction, wind and gravity g. The wind is described by the term on the right-hand side and depends on the cross sectional flow width b, wind velocity w, wind direction F, and the direction of the channel axis j. The non-linear term is a resistance term depending on gravity, discharge, hydraulic radius R, flow area A, and the coefficient of De Chézy, Ch. The parameter a in the advection term is the Boussinesq velocity distribution coefficient, which corrects for the non-uniformity of the velocity distribution. Conservation of momentum can now be written as:

$$\frac{\partial Q}{\partial t} + gA\frac{\partial H}{\partial x} + \frac{\partial}{\partial x}\left(\frac{\alpha Q^2}{A}\right) + \frac{g|Q|Q}{Ch^2 AR} = b\gamma w^2 \cos(\Phi - \varphi) \tag{6}$$

The flow area A is directly related to the area Ar used in the groundwater model. When the cross-section is fixed between two nodes then the area Ar is equal to A multiplied by the length of the channel section covered by the cell. In other cases the area Ar needs to be calculated using an averaging formula. The equations are discretised in space and time and solved using a four point implicit Preismann scheme, DUFLOW (1992).

Since DUFLOW assumes an impermeable bottom, the connection with the groundwater has to be made artificially. DUFLOW describes the channel using a series of sections connected by nodes. At the internal junctions, implicit boundary conditions assume that, in the absence of structures, the water level is continuous over such a junction node. Further, the flows $Q_{j,i}$, from node j, and additional flow q_i, into node i equal the flows from the node, such that there is no storage at the nodes, described by:

$$\sum_j Q_{j,i} + q_i = 0 \tag{7}$$

Equation (7) gives the opportunity to prescribe additional flow q_i to node i. This additional flow in DUFLOW is used to describe rainfall and the inflow from a drainage area. This term is also used to facilitate the link with the groundwater model.

4 INTERACTION

As we showed earlier, both DUFLOW and MODFLOW have the possibility to allow for flow from outside their own domain; i.e. DUFLOW can allow for groundwater discharge, which may be negative, into its channel system and MODFLOW can allow for the presence of a channel system. Important for the study of interaction is that the exchange of data is dynamic. By dynamic we mean that at regular intervals either of the programs receives accurate data about the head in the channel system and the groundwater discharge.

4.1 *Grid*

As we are working with discretised equations, we have defined grids for each of the models. Since the groundwater model is three dimensional and the surface water model one dimensional, it is not possible to match these grids exactly. Instead we want to look at the behaviour of the flows for a best match of both grids.

4.1.1 *Groundwater*
The groundwater model MODFLOW requires a head, or a water level for each of the

cells that contains a river reach. For each cell (i, j, k) it therefore needs one number, representing the average water level of the river reach in that cell. The easiest way to accomplish that is to place the nodes of the network in the centre of the cells. Consequently the head H_m at node m then equals the head at cell (i, j, k), giving:

$$H_{i,j,k} = H_m \tag{8}$$

Alternatively, larger sections can be defined. This means that there is not a one to one relation between a cell and a node. In this case an averaging formula is needed to calculate the average water level in a cell, resulting in:

$$H_{i,j,k} = Av(H_m, H_n) \tag{9}$$

The averaging formula Av that is used depends on the heads at nodes n and m located at either side of the section that is contained in cell (i, j, k). The formula may be weighted, depending upon additional knowledge, and therefore an explicit formula is not given here. Both methods can be used; however, each has its advantages and disadvantages. The method described by Equation (8) requires less administration but for larger problems it is easy to cause stability problems in the surface water calculations due to the small size of the sections and relatively large time steps that are used. The method described by Equation (9) leaves the size of the sections intact, but requires a large amount of extra administration. When using a Geographic Information System as the user interface the administrative work can be done in advance when setting up the model. Tables can be created for each cell (i, j, k) containing information for the averaging formula, all depending on the underlying geography. The simplest variant of the method using Equation (9) is to create a table indicating for each cell (i, j, k) at what node the water level should be used for calculating the interflow between groundwater and surface water. The table should, for example, indicate that cells (7,11-13,1) use the water level at node 14 and that cells (6,13-16,1) use the water level at node 15 (see Fig. 4).

4.1.2 *Surface water*
DUFLOW requires the knowledge of additional flow at its nodes. In correspondence with Equation (8), a one-to-one relation can be defined for the discharge term used by DUFLOW. The additional flow to node m, at cell (i, j, k), is simply given by:

$$q_m = q_{i,j,k} \tag{10}$$

The use of an averaging formula, such as in Equation (9), is unrealistic when no additional information is used for this formula. Important for the use of such a formula is that the flow direction of the surface water is known. In the Netherlands this assumption is not very realistic. Using an averaging formula without enough information would not necessarily lead to less accurate results but will require unnecessary effort both from the computer as well as the developer of the model. It is therefore suggested that the simplest form of averaging be used; for example, the methods with tables. These tables should now indicate to which node of the surface water model each of the leak terms calculated by the groundwater model contribute. This table should, for example, consists of a list of cells (7,11-13,1) and (6,13-16,1) that all contribute to the additional flow at node 15 (see Fig. 4).

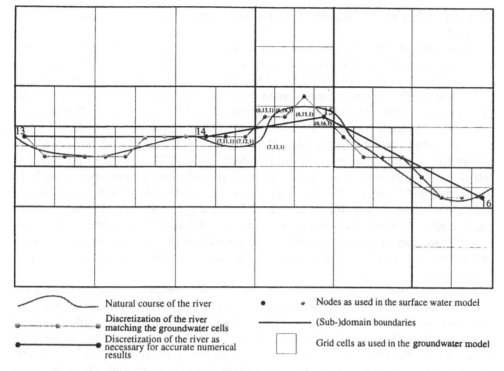

Natural course of the river

Nodes as used in the surface water model

Discretization of the river matching the groundwater cells

(Sub-)domain boundaries

Discretization of the river as necessary for accurate numerical results

Grid cells as used in the groundwater model

Figure 4. Discretization of a river reach on a groundwater grid.

4.1.3 *Discretisation*

Figure 4 concisely summarises the problems and questions with respect to the relation between scales and grids. It clearly divides the domain into sub-domains, each of them having grid sizes corresponding to the scales of the physical processes. For example, near the stream the scales are smaller and so are the grid sizes for the groundwater model. The natural course of the stream can be discretised in two ways as indicated in Figure 4. When the light grey method is used it matches the discretisation of the groundwater problem and Equations (8) and (10) are to be used for transfer of the interflow terms.

5 DATA INTERFACE

When the concept of decomposition is adopted such that the model that solves the overall problem is divided into several sub-models, then it is necessary to find a solution for the communication between the various sub-models. Ryan & Sieh (1993) take a data-centred approach for this. They show that such a data centred approach is not only suitable for linkage between sub-models (modelling applications) but serves also as a concept for the communication between the modelling application and other applications such as data-, GIS- and analysis applications. Important is that the data management interface (DMI) that facilitates the communication between the application and the database is not hard coded into the applications. This is to ensure flexibility when models are transferred to other platforms or when other databases are used.

In the case of coupling a groundwater and surface water model we followed this data-centred approach. A DMI was chosen so that the files were written in a format that enabled them to be transferred to other platforms without conversion (platform independent). This facilitated transfers in a computer network. Following current developments in data storage formats, we adopted the Network Common Data Format (NetCDF), (Jenter & Signell, 1992; Rew et al., 1993), developed by Unidata, which has an interface with the Hierarchical Data Format (HDF) developed at the National Centre for Super-computing Applications (NCSA). NetCDF is a data format that is self-describing, network-transparent and platform-independent and offers interfaces for Fortran and C programs. This means that data stored in NetCDF format is accessible over a network, which may consist of different types of computers using different operating systems. The files contain enough information about the data, so that multiple programs can use it, without prior knowledge about the dimensions or type of the data. It allows both Fortran and C programs to access the same data files in a similar way.

Jenter & Signell already mention that numerical modellers in hydrodynamics often use different computer platforms, which means that there is the need for a platform independent way to store the data. Further, these users often need subsets of large data files, which means that the data stored in the files should be available by direct access. Small subsets need to be extracted easily from large result files. The files should also contain information describing the data in enough detail that they can be distinguished from other similar files. Examples of this information, often called meta-data, that can be stored in the NetCDF files are: units, valid range, long names, data source, etc.

NetCDF is platform independent and data in the NetCDF format can be transmitted over computer networks without data modification because these files are encoded in XDR (Sun Microsystems, 1988). XDR is a non-proprietary, binary, external data representation developed by Sun Microsystems Incorporated and is implemented on many platforms such as Macintoshes, PCs, many workstation brands and main frames. This ensures that data stored in this format can be used widely. The user, however, does not need to know that NetCDF is using the complexity of the XDR file structure. The data is accessed using simple Fortran or C subroutines.

For programs to communicate, such as for the integration we propose between MODFLOW and DUFLOW, then a data format is needed that can be understood by both programs. Existing programs, however, write and read the data using their own defined format. To create communication, we need to deliver the data in the required format to each program and similarly, we have to read the data using this prescribed format. To avoid changing source codes, or if the source code is not available, then translation programs are required to facilitate communication between existing programs. However, the use of common data formats, such as NetCDF, makes things more flexible. The only requirement for using common data formats is that the read and write statements in the source code, of all programs intercommunicating, need to be changed to call up a library of subroutines. These subroutines will take care of the further data storage in the specified file. An additional advantage in the use of common data formats is that these clarify the source code with respect to the input and output routines.

Ultimately, when more models are using data stored in a common data format, loosely coupled models can evolve into a decision support system, as will be discussed in the next section. Of course, a common data format is not a requirement to make a decision support system, nor is it a requirement for a system of loosely coupled models; however, adoption of such a format makes the system very flexible and models can be added without too much effort. Both NCSA and Unidata are constantly updating their interfaces and the wide use of both formats give confidence for the support of these formats in the future.

6   DECISION SUPPORT SYSTEMS

Figure 5 shows the interaction between a local water resources system, the water manager (decision maker) and a decision support system. The decision support system here is given on an abstract level, which means that for many applications some of the features may be superfluous, but we aim to envelope all decision support systems for water management within this concept.

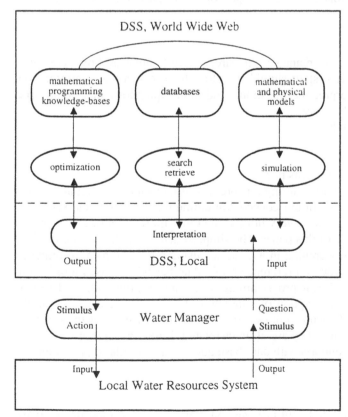

Figure 5. A decision support system for water management.

6.1 *Water manager*

The water manager or decision maker will in most cases not be just one person, but in fact can be a complex organisation or a system in which it is not readily clear who contributes to what part of the decision (Schultz, 1989). In general, however, a water manager will react to the state of the local water resources system for which he is responsible. Output from this system, e.g. reduced water quality, increasing population, flooding and so on, will stimulate him to act. He can do that directly by taking actions, which in turn will act as inputs to the system. These inputs will trigger people to be happy, satisfied, and angry, and the physical parameters of the system will obviously be changed. Hopefully the changes in people's attitudes and the physical parameters will be satisfactory, or at least predictable.

There are many cases when a water manager, whether a person or an organisation, would like to support his decision with extra knowledge. A water manager could ask his researchers to write a report on the matter and (partly) base his decision upon that, or these days he might ask his researchers to carry out some computer simulations. We have to keep in mind that automated decision making is still far from feasible and may never become real. The decision making process always involves multiple objectives, of which not all will be quantifiable. Computers are typically not able to simulate processes that are not quantifiable. For that reason, we should restrict our emphasis on decision support in the sense of advising rather than trying to create computer programs that can take over part of the decision making process.

6.2 *Interpretation*

The questions posed by a water manager will rarely be fully answered by a computer system. A decision support system should therefore have a level of interpretation, where people have to interpret the question. Additional questions that could arise during the interpretation phase are:
  – How much time do we have?
  – What data do we need?
  – Where do we get the data?
  – How accurate should the results be?

Many other questions will arise during this process of interpretation, but all of them eventually will lead to actions. In our concept we define three possible classes of action; two of them will be discussed very briefly, the other more elaborately because it is closely related to the topic of this paper. The simplest action can be a search and retrieve action. Examples are searching in a database, using a geographical information system or even browsing through a reference book. Another widely used tool is optimization; examples include linear programming, dynamic programming and genetic algorithms. Thirdly there is the option of simulation, which will be discussed in the next paragraph. It is very difficult to draw strict boundaries for each of these actions and it is not necessary, because they are all inter-related. Linear programming, for example, is often used in conjunction with data-base queries and simulation programs.

6.3 *Simulation*

Both of the simulation models DUFLOW and MODFLOW described in this paper are suitable to fit in the simulation part of a decision support system (DSS) for water management. In this way they represent all other hydrological computer models. The possibility of integration of both models makes them more valuable to the user of the DSS. If the DSS user only needs information about either groundwater or surface water, loose coupling has the advantage of a lesser overhead in data requirements and simulation time than a tightly coupled model. If the interest is on the interaction between groundwater and surface water the user has the advantage of using two familiar models that can communicate in a loosely coupled system. If a project has a generous time allowance, the DSS user might even decide to make a contribution to the system and add a model to the system that can communicate with the existing part of the DSS through NetCDF files, as mentioned earlier. When databases and optimisation programs use the same data format, they might even work together. The user would then only have to decide which programs he would like to integrate into a personalised loosely coupled system, which would involve some modest programming, and would have the benefits of interacting models.

6.4 *Distributed DSS*

It is not necessary to have all the computer power located in a central place. It is more advantageous to have data and models at the computer where they will be updated. Current technologies allow data searches through a network. We believe that in the near future data communication will be less of a problem than it is now, and that there is no need to have all data and models locally. Instead we can fetch the data and the models when we need them, or if data communication advances exceed expectations, we could run models on remote computers where each of these models will access data over a network. This approach termed 'World Wide Web', describes a specific way to search for and retrieve information, but in this case is used as a general term for the location of information.

7 CONCLUSIONS

The integration of two models for studying the interaction of groundwater and surface water is part of a research project to develop a concept for a decision support system that can be used by water managers. We showed that loose coupling of models allows us to be flexible in adding other features or hydrological models to this system. Advances in computer technologies will probably allow us to implement a distributed system, as described, on the scale of the Netherlands, within a couple of years.

We also showed that coupling of models describing different hydrological processes is not a trivial task. Major concerns are on scale issues and the choice of the grid. We saw that for the choice of the grid the underlying hydrology may play a role and, further, that grid choice also relates to the spatial scales of the processes. Grid issues particularly arise at the interfaces of sub-domains and at the groundwater-surface water boundary. The user of a loosely coupled system should not be bothered by the choice of a grid at these interfaces and actual tests should determine what is the better choice.

The division into sub-domains is highly dependent on the hydrological processes involved and should be left to the user.

The use of common data formats, like NetCDF, is necessary to make the integrated system of models flexible and therefore suitable to fit in a distributed DSS.

REFERENCES

Abbott, M.B. 1979. *Computational Hydraulics*. London, Pitman.

DUFLOW. 1992. *A Micro-Computer Package for the Simulation of One-Dimensional Unsteady Flow and Water Quality in Open Channel Systems.* Manual for DUFLOW version 2.00, ICIM, Rijswijk.

Geldof, G.D. 1993. Enquête over gebruik modellen in het waterbeheer, pp 13-23. In: G.D. Geldof & L.R. Wentholt (eds), *Modelleren Op Maat.* STOWA-rapporten nr 9, Utrecht.

Haagsma, I.G. 1995. The integration of computer models and data bases into a decision support system for water management, pp. 253-261. In: Modelling and Management of Sustainable Basin-Scale Water Resource Systems. *IAHS Symposium H6, Boulder, USA, 3-14 July 1995, Boulder. IAHS Publication No 231*, Slobodan P. Simonovic, Zbigniew Kundzewicz, Dan Rosbjerg and Kuniyoshi Takeuchi (eds), IAHS, Wallingford.

Jenter, H.L. & Signell, R.P. 1992. NetCDF: A public-domain solution to data-access problems for numerical modelers, pp. 77-82. In: Estuarine and Coastal Modeling. *Proceedings of the 2nd International Conference, Tampa, 13-15 November 1992*, M.L. Spaulding, K. Bedford & A. Blumberg (eds), ASCE, New York.

McDondald, M.G. & Harbaugh, A.W. 1983. *A Modular Three-Dimensional Finite-Difference Ground-Water Flow Model.* US Geological Survey, Open-File Report 83-875.

Ngo, X.T. 1994. *Koppeling tussen Grondwatermodel en Oppervlaktewatermodel MODFLOW.* Report Waterleiding Maatschappij Overijssel N.V.

Rew, R; Davis, G. & Emmerson, S. 1993. *NetCDF Users's Guide. An Interface for Data Access, Version 2.3*, Unidata.

Ryan, T.P. & Sieh, D. 1993. Integrating hydrologic models, geographic information systems and multiple databases: A data centered approach, pp. 26-34. In: Proceedings of the Fedral Interagency Workshop on Hydrologic Modeling Demands for the 90's. US Geological Survey Water-Resources Investigation Report 93-4018, USGS.

Schultz, G.A. 1989. Ivory tower versus ghosts? – or – the interdependency between systems analysts and real-world decision makers in water management, pp 23-32, In: D.P. Loucks (ed.), *Closing the Gap Between Theory and Practice.* IAHS Publication No 180. Wallingford, IAHS.

Sun Microsystems. 1988. *External Data Representation Standard: Protocol Specification*, RFC 1014. Information Sciences Institute.

Yen, B.C. & Riggins, R. 1991. Time scales for surface-subsurface modeling, pp. 351-358. In: *Proceedings of the 1991 National Conference on Irrigation and Drainage*, Honolulu, 22-26 July 1991. New York, New York.

CHAPTER 13

Coupled air-water flow and contaminant transport in the unsaturated zone: A numerical model

PHILIP BINNING
Department of Civil Engineering and Surveying, The University of Newcastle, Australia

MICHAEL A. CELIA
Water Resources Program, Department of Civil Engineering, Princeton University, USA

ABSTRACT: A new two-dimensional model of multiphase flow and contaminant transport in the unsaturated zone is presented. The flow model incorporates movement of both the air and water phases in heterogeneous materials with spatial and time varying boundary conditions. The transport model includes retardation, reactions, and both equilibrium and non-equilibrium mass transfer between the air and water phases. The equations describing these phenomena are solved using a finite element method in two-dimensional cartesian and radial coordinates. This coupled model of the air and water phases can be used to model a wide variety of phenomena. These include soil venting to remove volatile organic waste, waste disposal site design, and gas exchange with the atmosphere for use in climate models. The importance of the air phase is illustrated through the discussion of some of these examples.

1 INTRODUCTION

The unsaturated zone is a multiphase system, consisting of at least three phases: A solid phase of the soil matrix, a gaseous phase, and the water phase. Additional phases may also be present, for example a separate phase organic liquid or an ice phase. The traditional approach of hydrologists to study the unsaturated zone has been to focus exclusively on the water phase (Richards, 1931). Recently interest has grown in a number of problems where the other phases in the unsaturated zone may be important. These include: Evaluation of remediation technologies such as soil venting (Johnson, 1990), improving estimates of water infiltration by accounting for the effect of the air phase on the movement of water (Touma & Vauclin, 1986), measuring carbon dioxide movement from the soil to the atmosphere for use in climate models, improving estimates of evaporative flux from soils, and evaluating design criteria for volatile hazardous waste disposal. Each of these examples has the common feature that the gas phase plays an important role in its description.

While there have been many studies of water movement in the unsaturated zone, few have focussed on the movement of the air phase. Of those that have, most have ignored the influence of water movement on the air phase. In this paper a coupled air and water flow and contaminant transport model is developed.

The model that is developed is a mathematical model based on the established physics of porous media flow and transport. The governing equations form a system of highly non-linear, coupled, partial differential equations. The governing equations can be divided into two sets of equations, the first describing the fluid flow and the second contaminant transport. These equations can rarely be solved analytically for situations of practical importance and so numerical solutions must be developed. A numerical solution to these equations is described here and some examples are discussed to demonstrate the performance of the numerical method and to illustrate the physics behind the equations.

2   MODEL DESCRIPTION

The introduction outlined a broad range of issues that are of interest in the unsaturated zone. These included soil venting, soil gas exchange with the atmosphere, rainfall-runoff estimation and hazardous waste disposal. It was apparent from these problems that there is a need for a model of soil processes that includes:

1. Coupled water and air phase flow,
2. Contaminant transport in both phases.

In this section the two-dimensional equations describing each of these processes are outlined.

The first set of equations describes the flow of the air and water phases in the porous medium. The equations are conservation of mass statements in each fluid with the fluid flux expressed using Darcy's law:

$$q_\alpha = -K_\alpha \left(\nabla h_\alpha - \frac{\rho_a}{\rho_{ow}} (\sin \psi) i_z \right) \tag{1}$$

The mass balance equations assume that the water is slightly compressible and that the air phase is fully compressible with the air density linearly dependent upon the air pressure.

$$\frac{\partial \theta_w}{\partial t} + S_{sw} \frac{\theta_w}{\phi} \frac{\partial h_w}{\partial t} - \nabla(K_w(\nabla h_w - i_z \sin \psi)) = F_w \tag{2}$$

$$(\phi - \theta_w)\frac{\rho_{oa}}{h_{oa}}\frac{\partial h_a}{\partial t} - \rho_a \frac{\partial \theta_w}{\partial t} - \nabla\left((\rho_a K_a)\left(\nabla h_a - \frac{\rho_a}{\rho_{ow}} i_z \sin \psi \right) \right) = F_a \tag{3}$$

In these equations q_α is the flux of phase α, with α denoting the fluid phase. In this work α will refer to air and water. K_α is the conductivity tensor of phase α, h_α the pressure head of the fluid (pressure in equivalent water column height), and ρ_α is the density of phase α. The water is assumed to be slightly compressible and the air phase assumed to be a linear function of air pressure given by (Touma & Vauclin, 1986):

$$\rho_a = \rho_{oa}\left(1 + \frac{h_a}{h_{oa}}\right) \tag{4}$$

where ρ_{oa} is the density of air at standard temperature and pressure h_{oa}, i_z is the unit normal in the vertical direction, ϕ is the porosity, θ_α is the fluid content (volume of fluid/total volume), S_{sw} is the specific storativity of the water phase accounting for the elasticity of the porous medium and the slight compressibility of the water phase, ψ is the angle of the z coordinate to the horizontal, and F_α is a source sink term in phase α.

These equations are closed through the specification of constitutive relations for the conductivity and the moisture content. These constitutive relations can have many forms. Some of the most commonly used are those of Van Genuchten (1980) and Parker (1987), which specify the relationship between the capillary pressure $h_c = h_a - h_w$ and the moisture content, and the conductivities and moisture content. A complete statement of the flow equations also includes the initial conditions and the specification of the boundary conditions as either a fixed pressure head or fluid flux at each point on the boundary.

The second set of equations describe the transport of pollutants in either phase. The transport processes are represented with the advection-dispersion-reaction equation which is given by:

$$\frac{\partial(\theta_\alpha c_\alpha)}{\partial t} + \nabla(q_\alpha c_\alpha) - \nabla(\theta_\alpha D_\alpha \nabla c_\alpha) + \lambda_\alpha \theta_\alpha c_\alpha = R_\alpha$$

In the transport equation c_α is the concentration of contaminant in each phase, λ_α is the reaction rate of the contaminant and R_α are source sink terms in either phase. The dispersion process is assumed Fickian with a dispersion tensor given by:

$$D_{\alpha_{ij}} = \alpha_{T_\alpha}|v_\alpha|\delta_{ij} + \left(\alpha_{L_\alpha} + \alpha_{T_\alpha}\right)\frac{v_{\alpha_i} v_{\alpha_j}}{|v|} + D_{m_\alpha}\delta_{ij} \tag{5}$$

with α_{L_α}, α_{T_α} the longitudinal and transverse dispersivities respectively, $v_\alpha = q_\alpha/\theta_\alpha$ the fluid velocity, δ_{ij} the Kronecker Delta function and D_{m_α} the molecular diffusion in phase α.

Many types of pollutants are volatile, partitioning between the water and gas phases. Other pollutants adsorb onto the solid phase. In both cases a description of the partitioning of the contaminant between phases is necessary. Partitioning of the contaminant between the solid and liquid phases is described through a linear sorption isotherm. For the air and water phases the most common form of partitioning is to assume equilibrium partitioning governed by Henry's Law which states that the concentration of contaminant in each of the phases remain at a fixed ratio given by the Henry's constant $H = c_a/c_w$ The more general case of kinetic mass transfer has been modelled in one dimension by Binning (1994).

For the case of equilibrium partioning between the water and air phases, the air phase concentration can be specified in terms of the water phase concentration and the air and water equations added to find a single equation in c_w:

$$\frac{\partial((R\theta_w + H\theta_a)c_w)}{\partial t} + \nabla((q_w + Hq_a)c_w) - \nabla((\theta_w D_w + \theta_a HD_a)$$

$$\nabla c_w) + (\lambda_w \theta_w + H\lambda_a \theta_a)c_w = F_w + F_a \tag{6}$$

In the above equation R is the retardation coefficient for sorption of contaminant onto the solid phase which is defined as:

$$R = 1 + \frac{\theta_s \rho_s k_d}{\theta_w} \tag{7}$$

with θ_s, ρ_s the volumetric content and density of the solid phase and k_d the distribution coefficient. The form of the retardation equation is determined by the assumption that the ratio of solid and water concentrations is a constant k_d.

Initial conditions must also be specified for the transport equation and boundary conditions are given at each point on the boundary as either a fixed concentration, dispersive flux or total boundary flux.

Fully heterogeneous material definitions are included in the model through the definition of the constitutive relations and the transport parameters. The equations have been solved in both cartesian and radial geometries (see Binning, 1994); however, results will only be given here for problems with a cartesian geometry.

3 NUMERICAL SOLUTION

The numerical method for the solution of the flow equation has been published for the one-dimensional case by Celia & Binning (1992a) and is based on the work of Celia et al. (1990). Celia et al. solved Richards' equation using a method that employs the mixed form of the equations to guarantee mass conservation, modified Picard iteration to linearize the equations and lumped finite elements. The two-dimensional method is analogous and is presented here.

In the equations given above, the flow of the fluids is assumed to be independent of the concentration of contaminant in the fluid. This is true where concentrations are small, however for higher concentrations arising in some problems (see for example Mendoza et al., 1990) the density of the fluid phase may be strongly influenced by contaminant concentration which in turn affects the flow of the fluid.

Since the flow equation is independent of contaminant concentration, the numerical solution of the governing equation can proceed in two steps. The flow equation is solved first for the pressures in each phase, the flow field determined using Darcy's law, and then the transport solved for the concentration in the water phase.

The flow equations are nonlinear because of the dependence of the coefficients on the solution itself (e.g. $K_\alpha = K_\alpha(\theta_w) = K_\alpha(h_a - h_w)$), and coupled through the dependence of the constitutive relations on the capillary pressure. Each of these mathematical difficulties must be dealt with by the numerical method. The numerical method is outlined here. For further details the reader is referred to Binning (1994) and Binning et al. (1995).

The equations are discretized in time using an Euler backward finite difference scheme which ensures the stability of the resultant numerical method. Linearization of the equations is achieved using a Picard iteration scheme which lags the nonlinear coefficients one iteration behind the solution. At each time step the equations are solved and the nonlinear coefficients updated. Convergence is reached when the difference between solution at the old and new iteration is small. The equations are expressed in mixed form, that is with the temporal derivatives applied to the moisture content function and the spatial derivatives applied to the pressure head. The mixed form of the equations have the advantage that the resultant numerical approximation is fully mass conservative. This result was demonstrated by Milly (1985), Redinger et al. (1984) and others for Richards' equation. The spatial discretization employs Galerkin finite elements with bilinear basis functions on a regular rectangular grid. Mass lumping is employed to control the oscillations that typically arise in finite element methods.

Variable time stepping is also used. When the gradient of pressure head is large the nonlinearities in the equation are strong and small time steps must be taken to guarantee convergence of the iteration scheme. As the pressure front becomes more smooth, the nonlinearities weaken and larger time steps can be used. The algebraic equations are assembled in a matrix with both the air and water equations combined in a single matrix. By alternating the equations for air and water in the matrix the band width of the matrix can be minimized allowing the use of efficient numerical techniques that take advantage of the sparse matrix structure. Finally the algebraic equations resulting from the numerical discretization are solved using a banded gaussian elimination matrix solver. More efficient matrix methods are available to solve these equations (e.g. iterative methods), but the simple technique used here is sufficient for the small problems considered here.

Once the individual phase pressures are known the mass flux of each phase can be calculated using Darcy's law. In this case a standard centred difference approximation is used to discretize Darcy's law and find the flux at the centre of each element.

The numerical method used to solve the contaminant transport equation is an Euler backward difference scheme in time and a Galerkin finite element scheme in space with bilinear basis and test functions defined on the same grid as was used for the solution of the flow equation. Lumping of the time terms is used to control oscillations arising from the finite element solution. Note that lumping has been shown to be ineffective in one-dimension (Daus, 1985). However in higher dimensions Binning (1994) demonstrated that lumping is an effective strategy for controlling finite element oscillations.

4 EXAMPLE: MOVEMENT OF AIR AND WATER UNDER INFILTRATION

The first example considers infiltration from a strip source into an initially dry uniform Touma & Vauclin (1986) coarse sand. The boundary conditions, initial conditions and the discretization used for this simulation are illustrated in Figure 1.

Variable time step sizes were used to solve this problem. Initially, when the infiltrating pressure front is steep, the nonlinearities are strong and a small time step must be chosen. In this case the initial time step size was 100 seconds. As time passes the infiltrating front smooths and the nonlinearities weaken. This can be seen as the Picard iteration scheme requires fewer iterations to converge as time passes. As the

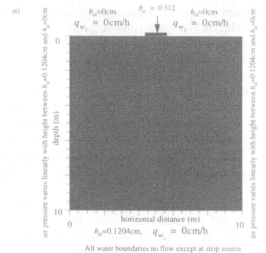

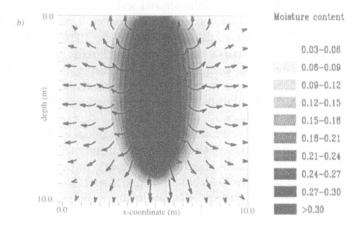

Figure 1a) Material distribution, boundary and initial conditions for infiltration from a strip source into an initially dry transect of coarse Touma & Vauclin (1986) sand, b) Flow solution at time 25 hours for the boundary conditions shown in (a). The figure shows water infiltration in grey scale and has superposed the paths of particles moving from their initial position to that at $t = 25$ h.

amount of iterations decreases below 5 for the current time step, the time step size is increased by a factor of 1.05. The final time step size for this simulation was 167 seconds. The amount of time-step acceleration possible for this two-dimensional problem is small compared to those observed in one-dimensional simulations (Celia & Binning, 1992). In one dimension it was possible to increase the time-step to as much as 37 times its original value for some problems. The small increase in time step size in the two-dimensional examples reflects the more difficult nature of two-dimensional problems. With the Picard iteration scheme, there is always a trade-off between the number of iterations and the size of time step that can be taken, with larger time steps generally requiring more iterations.

The flow of air and water are shown in Figure 1. The grey scale in Figure 1 shows the distribution of water at time 25 hours. The darker shades indicate higher moisture contents and the lighter shades lower moisture contents. The paths of the air are shown by arrows on the figure. Each arrow represents the trajectory of an air particle over a 25 hour period. The figure shows the expected bulb-like infiltration of water into the unsaturated zone with capillary diffusion in the vertical and horizontal directions. The air paths radiate outwards as the air is expelled by the infiltrating water. Interestingly, with the water moving downwards and the air free to escape through all sides of the transect, it moves upwards and sideways towards the boundary as well as downwards toward the bottom. This occurs because there is a slight pressure increase in front of the infiltrating water which causes a pressure gradient to occur towards all the boundaries of the domain that are open to the air phase.

The water and air flow in a uniform material can be fairly well predicted by a one-dimensional model. However, most field examples are not uniform. A second example considers the effect of heterogeneity on the flow of air and water. The domain and boundary conditions are the same as those in Figure 1, with loam lenses embedded in the homogeneous sand. The lenses are a hypothetical loam with properties described by the van Genuchten (1980) pressure saturation relation:

$$\theta_w = \frac{\theta_{ws} - \theta_{wr}}{\left[1 + |\alpha h_c|^n\right]^{1-\frac{1}{n}}} + \theta_{wr}$$

with parameters $\theta_{ws} = 0.35$ cm^3/cm^3, $\theta_{wr} = 0.1$ cm^3/cm^3, $\alpha = 0.034$ cm^{-1}, $n = 3.2$ and the Parker (1987) conductivity functions with a saturated water conductivity of $K_{ws} = 1.512$ cm/h and a saturated air conductivity of $K_{as} = \mu_w/\mu_a (K_{ws}) = 2.79$ cm/h.

The distribution of the loam lenses is illustrated in Figure 2a and the flow of air and water at time 25 hours is also shown in Figure 2b. In this example the loam lenses are essentially impermeable units embedded in the coarse sand matrix. The water and air flow around the lenses rather than through them. For example, rather than flowing through lenses 2 and 4 the water flows on top of the lenses and around them.

An important issue that often arises in the discussion of heterogeneity is that of scale. At the scale of the example above, the flow of water is dominated by the advection of the infiltrating water. There is little spreading of the water through the action of capillary forces. There is a large contrast in conductivity between the sand and loam so that the water preferentially moves through the sand. A third example is used to investigate the effect of the scaling of heterogeneities. The example is 2 × 2 m with the same configuration of lenses and the same material properties as were used in the previous example. The strip source at the top of the domain is 1/5 the size of the source at the top of the larger domain. The flow solution is shown in Figure 3.

At the smaller scale the effects of capillary diffusion are far more prevalent. In this example the actual amount of capillary diffusion acting in a given time is the same as at the larger scale. However, relative to the scale of the heterogeneities, the depth of infiltration and the shorter time scale, the action of capillary forces are more noticeable with the water spreading and forming a plume that is much wider relative to the width of the strip source on the soil surface. This enhanced spreading is par-

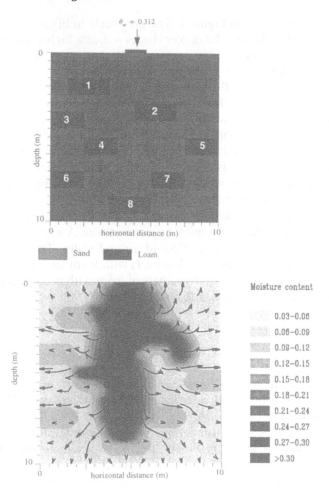

Figure 2a) Material distribution for infiltration from a strip source into a transect of coarse Touma & Vauclin (1986) sand embedded with loam lenses using the boundary and initial conditions shown in Figure 1. The numbering of the lenses is used in the discussion in the text. b) Water distribution and air movement after 25 hours of simulation. The distribution of material properties is shown in (a).

ticularly notable when the water encounters lenses. At the smaller scale the movement into the lenses is more pronounced than at a large scale. For example, when the water encounters lens 1 (Fig. 3) it moves into the lens. However, the advective movement that was present at the larger scale is still present at the small scale. For example, when the water encounters lens 2, the lens saturates with water through capillary action, but the advection still preferentially occurs in the higher conductivity material and the water flows above lens 2 and around it rather than passing through it.

These two figures contain at least three different scales: The scale of the heterogeneities, the time scale, and the scale of the domain and water source on the upper

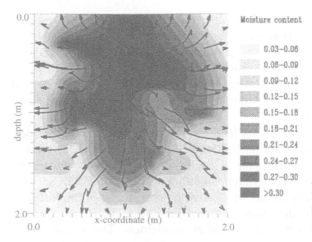

Figure 3. Simulation as in Figure 2 at a smaller spatial scale with a domain 2 × 2 m having the same configuration of lenses as in Figure 2. The solution is shown at $t = 8.3$ hours.

boundary. Separating the effects of these scales is difficult. For example, even if the larger scale simulation was run for a shorter time, the solution is unlikely to appear similar to the small scale solution. The governing equations are highly nonlinear, making scaling from one solution to another difficult. The limited examples discussed here cannot adequately address the scaling issues that have been raised. However, they do raise interesting questions about the way that heterogeneities affect the balance between the forces that shape the resultant solution.

In the context of remediation of a volatile organic compound, heterogeneities may play a significant role. It can be envisaged that large amounts of a volatile organic contaminant may be found embedded in the loam lenses. Yet when air is pumped through the unsaturated zone in a venting scheme, the air is likely to flow around the lenses and avoid the contaminant. In practice, it is observed at field sites that when the venting scheme is in operation the contaminant concentration decreases. After the pump is turned off the contaminant concentration in the air phase is often observed to recover. This could be because the contaminant that is removed by the venting operation is predominantly in the high permeability units. When the air pumping is halted, the volatile contaminant diffuses from the low permeability units back into the high permeability materials, increasing the concentration there.

5 CONCLUSIONS

The model of air and water flow and contaminant transport discussed in this paper has been applied to a number of additional problems, including the simulation of low level radioactive waste disposal (Binning et al., 1995), the Cape Cod unsaturated zone field experiment of Rudolph et al. (1993) and the simulation of radial flow of air towards a well arising in a soil venting scheme. The model is easily adapted to a wide range of problems. All types of boundary conditions can be specified at each individual boundary node. Heterogeneities can be incorporated by assigning individual material properties to each element in the domain or to areas of the domain. The model has been found to be robust and accurate for all problems tested.

Features of the numerical method used to solve the governing equations include guaranteed mass balance, no oscillations in the finite element solution, and the ability to handle the most difficult problems arising in situations with steep infiltrating pressure fronts. It should be noted, however, that because of the strong nonlinearity of the governing equations, a fine spatial discretization and small time steps are needed to guarantee convergence of the time iteration scheme in problems where sharp changes in moisture content occur.

The model has been used in this work to demonstrate the importance of considering heterogeneity when modelling water infiltration and air movement in the unsaturated zone. It has been shown that because of the nonlinearities in the governing equations, it is not possible to scale up solutions in either space or time. The scale of the problem has been shown to determine the relative balance between the forces of capillary action and gravity that determine the resultant solution. Heterogeneities have been shown to greatly affect both water and air movement. The solutions presented here are for materials with highly contrasting material properties. However, even in cases with milder heterogeneity, previous work with Richards' equation (e.g. Bouloutas, 1990) has demonstrated the importance of considering material heterogeneity. If a detailed picture of the flow of either water or air is needed in the unsaturated zone, the effect of heterogeneity must be considered.

ACKNOWLEDGEMENTS

The authors would like to thank Howard Sweed and the Princeton group for their contributions to this work. This research was partially funded by the NSF under grant 8657419-CES and by the USNRC under contract number NRC-04-88-074. Although research has been funded in part by these organisations it is not subject to agency review and so no official endorsement should be inferred.

REFERENCES

Binning, P., Celia, M. A. & Johnson, J.C. 1995. Two-phase flow and contaminant transport in unsaturated soils with application to low-level radioactive waste disposal, Nuclear Regulatory Commission Report, NUREG/CR-6114 Vol. 2.

Binning, P. 1994. Modeling unsaturated zone flow and contaminant transport in the air and water phases, Ph.D. thesis, Department of Civil Engineering and Operations Research, Princeton University, Princeton, New Jersey.

Celia, M.A., Bouloutas, E.T. & Binning, P. 1990. Numerical methods for nonlinear flows in porous media. In: Gambolati et al. (eds), *Proc VIII Int. Conf. Computational Methods in Water Resources.* Venice, Italy.

Celia, M.A., Bouloutas, E.T. & Zarba, R.L. 1990. A general mass-conservative numerical solution for the unsaturated flow equation. *Water Resources Research* 26(7): 1483-1496.

Celia, M.A. & Binning, P .1992. Two-phase unsaturated flow: one dimensional simulation and air phase velocities. *Water Resources Research* 28(10): 2819-2828.

Daus, A.D., Frind, E.O. & Sudicky, E.A. 1985. Comparative error analysis in finite element formulations of the advection-dispersion equation. *Advances in Water Resources* 8: 86-95.

Johnson, P.C., Stanley, C.C., Kemblowski, M.W., Byers, D.L. & Colthart, J.D. 1990. A practical approach to the design, operation, and monitoring of in situ soil-venting systems. *Ground Water Management Review*: 159-178.

Mendoza, C.A. & Frind, E.O. 1990. Advective dispersive transport of dense organic vapours in the unsaturated zone 1. model development. *Water Resources Research* 26(3): 379-387.

Milly, P.C.D. 1985. A mass-conservative procedure for time stepping in models of unsaturated flow. *Advances in Water Resources.* 8: 32-36.

Parker, J.C., Lenhard, R.J. & Kuppusamy, T. 1987. A parametric model for constitutive properties governing multiphase flow in porous media. *Water Resources Research* 23(4): 618-624.

Redinger, G.J., Campbell, G.S. Saxton, K.E. & Papendick, R.I. 1984. Infiltration rate of slot mulches: Measurement and numerical simulation. *Soil Science Society of America Journal* 48: 982-986.

Richards, L.A. 1931. Capillary conduction of liquids in porous media. *Physics* 1: 318-333. New York.

Rudolph, D.L., Kachanoski, R.G., Celia, M.A. & LeBlanc, D.R. 1993. Infiltration and solute transport experiments in unsaturated sand and gravel, Cape Cod Massachusetts: Experimental design and overview of results. Submitted to *Water Resources Research*.

Touma, J. & Vauclin, M. 1986. Experimental and numerical analysis of two-phase infiltration in a partially saturated soil. *Transport in Porous Media* 1: 27-55.

Van Genuchten, M.Th. 1980. A closed form equation for predicting the hydraulic conductivity in soils. *Soil Sci. Soc. Am. J.* 44: 892-898.

CHAPTER 14

A Laplace transform numerical model for
contaminant transport in stratified porous media

J.C. WANG & L.W. APPERLEY
School of Civil & Mining Engineering, The University of Sydney, Australia

ABSTRACT: This paper presents a Laplace transform numerical technique for the description of 1D and 2D contaminant transport in saturated stratified porous media, which was verified through comparison with an analytic method and a finite element method. The technique provides a stable approach to solving the advection-dispersion equation describing solute transport in saturated porous media. No time-stepping is required in the Laplace transform numerical method. Thus, a solution can be obtained directly for any point in time.

1 INTRODUCTION

A conventional approach to the prediction of concentration profiles of contaminant transport is to solve the advection-dispersion equation. The analytical solutions are only available for certain idealized cases. Although numerical methods, such as finite difference methods and finite element methods, give reasonable results, the computational effort involved is quite large, particularly if it is necessary to calculate concentration profiles at large times. The treatment of the time derivative in the mass transport equation is usually identical in both the finite difference (FD) and finite element (FE) methods. The continuous time coordinate is discretized into time steps. It is generally well known that artificial oscillations in the solution can be caused by the improper selection of a time step in these methods.

Laplace transform numerical techniques have been used previously to eliminate the effects of time terms in solving the transient transport equation by Sudicky (1989) and Wang & Apperley (1994). The Laplace transform numerical method involves an application of the Laplace Transformation to the time-dependent partial differential equation. The time derivative is analytically removed via transformation and the transient transport equation becomes an ordinary differential equation with only spatial derivatives. Numerical inversion of the Laplace-transformed nodal concentration is performed using the Stehfest (1970) algorithm after solving the resulting transformed system of algebraic equations in Laplace space by the finite difference method. The ef-

169

fects of the traditional treatment of the time derivative on accuracy and stability are rendered irrelevant because time is no longer a consideration.

The Laplace transform numerical technique is used to model contaminant movement in layered or stratified porous media. Two problems are considered here. In the first, multiple layered media are considered with the flow perpendicular to the interfaces. In the second problem, the flow is parallel to the stratifications and the flow across the common boundary is assumed to be negligible. Sorption is neglected in both problems. A general Galerkin finite element method is used to demonstrate the model's accuracy for the first problem. On the other hand, analytical solutions of Al-Niami & Rushton (1979) and numerical solutions of Wang et al. (1986) are used and compared with the results obtained by the Laplace transform numerical method for the second problem.

2   THEORETICAL BACKGROUND

2.1   Governing equation

The mass transport equation through porous media in which advection, diffusion, dispersion and adsorption take place can be written as:

$$
\frac{\partial}{\partial x}\left[nD_x\frac{\partial C}{\partial x}\right]+\frac{\partial}{\partial y}\left[nD_y\frac{\partial C}{\partial y}\right]+\frac{\partial}{\partial z}\left[nD_z\frac{\partial C}{\partial z}\right]=
$$
$$
n\frac{\partial C}{\partial t}+\rho\frac{\partial s}{\partial t}+nV_x\frac{\partial C}{\partial x}+nV_y\frac{\partial C}{\partial x}+nV_z\frac{\partial C}{\partial z}
$$

(1)

where C is the concentration of dispersing mass, n is the porosity of the media, ρ is the media dry density, S is the concentration of the solute on the solid phase, D_x, D_y, D_z are hydrodynamic dispersion coefficients in the x, y, and z directions, and V_x, V_y, V_z are seepage velocities in the x, y, and z directions. In the simplest case, the sorption process can be modelled as being linear and reversible and so the mass of contaminant removed from solution is proportional to the concentration in the solute and can be written as $S = K_dC$, in which S is mass of solute removed from solution per unit mass of solid, K_d is the partition or distribution coefficient, and C is the concentration of solute (mass of solute per unit volume of fluid). If the soil medium is assumed to be incompressible, the porosity (n) and soil density (ρ) can be treated as constants. For a one-dimensional, homogeneous, isotropic medium with flow aligned with the x direction, in which the dispersion coefficient is considered to be a constant, Equation (1) becomes:

$$
D_x\frac{\partial^2 C}{\partial x^2}=\left[1+\frac{\rho K_d}{n}\right]\frac{\partial C}{\partial t}+v_x\frac{\partial C}{\partial x}
$$

(2)

Under two-dimensional conditions, the governing equation can be written as:

$$
D_x\frac{\partial^2 C}{\partial x^2}+D_y\frac{\partial^2 C}{\partial y^2}=\left[1+\frac{\rho K_d}{n}\right]\frac{\partial C}{\partial t}+v_x\frac{\partial C}{\partial x}+v_y\frac{\partial C}{\partial y}
$$

(3)

2.2 *Laplace transformation*

The Laplace transformation is applied to the governing equation and its associated initial and boundary equations before the spatial discretization step. For two-dimensional conditions, applying the Laplace transformation to the governing Equation (3) and making use of the basic operational property to eliminate the temporal derivative in Equation (3) leads to:

$$D_x \frac{\partial^2 \phi}{\partial x^2} + D_y \frac{\partial^2 \phi}{\partial y^2} = \left[1 + \frac{\rho K_d}{n} \right](s\phi - C_o) + v_x \frac{\partial \phi}{\partial x} + v_y \frac{\partial \phi}{\partial y} \tag{4}$$

in which C_0 is the initial concentration of solute (mass of solute per unit volume of fluid), s is the Laplace transform variable, $R = 1 + \rho K_d/n$ is commonly called the retardation coefficient, ϕ is the concentration of solute in Laplace space. Under one-dimensional conditions, the transformed mass transport equation is similar to Equation (4) but without D_y and V_y terms.

2.3 *The finite difference* (FD) *scheme in Laplace space*

In the finite difference approximation, derivatives are replaced by differences taken between nodal points. A block-centred grid method is used in the space discretization of the transformed mass transport Equation (4). If we consider an equal-interval grid of points, such that $\Delta x=$ constant and $\Delta y=$ constant, then the finite difference form of the governing equation in transformed space may be written as:

$$\begin{aligned}
\left[\frac{2D_x}{(\Delta x)^2} + \frac{2D_y}{(\Delta y)^2} + \left(1 + \frac{\rho K_d}{n} \right)s \right]\phi_{ij} &= \left[\frac{D_x}{(\Delta x)^2} + \frac{V_x}{2\Delta x} \right]\phi_{i-1,j} \\
+ \left[\frac{D_x}{(\Delta x)^2} - \frac{V_x}{2\Delta x} \right]\phi_{i+1,j} &+ \left[\frac{D_y}{(\Delta y)^2} + \frac{V_y}{2\Delta y} \right]\phi_{i,j-1} \\
+ \left[\frac{D_y}{(\Delta y)^2} - \frac{V_y}{2\Delta y} \right]\phi_{i,j+1} &+ \left(1 + \frac{\rho K_d}{n} \right)C_{o_{ij}}
\end{aligned} \tag{5}$$

2.4 *Numerical inversion of the Laplace transformation*

The FD approximation of the partial differential equations in the Laplace space results in a FD system of simultaneous Equations (5). The computational procedures for solving these simultaneous equations need arithmetic values of the s variable of the Laplace space. For a desired observation time t, they are provided by the first part of Stehfest's algorithm (1970) as:

$$s_I = \frac{\ln 2}{t} I \;\; ; \;\; I = 1 \dots NS \tag{6}$$

in which s is the Laplace transform variable, NS is the number of summation terms in the algorithm; NS must be even number. To obtain a solution at a time t, all vectors $\phi_{i,j}(I)$, $I = 1 \dots NS$, of those equations in Laplace space are needed. This means that the system of simultaneous equations has to be solved *NS* times. The vector of the un-known concentration *C* at any time is obtained by using the Stehfest (1970) algo-rithm to numerically invert the Laplace solutions, $\phi_{i,j}(I)$. The procedure is described by the following equations:

$$C(t) = \frac{\ln 2}{t} \sum_{I=1}^{NS} V(I)\phi_{i,j}(I) \tag{7}$$

where:

$$V(I) = (-1)^{(NS/2)+1} \cdot \sum_{k=(I+1)/2}^{\min(I,NS/2)} \left[\frac{k^{NS/2}(2k)!}{((NS/2)-k)!k!(k-1)!(I-k)!(2k-I)!} \right]$$

The solution in the Laplace space eliminates the stability and accuracy problems caused by the treatment of the time derivative in conventional FD approximations, thus allowing an unlimited time step size. The truncation error of the method is lim-ited to the truncation error caused by the space discretization because the domain is not discretized in time and the method provides a solution inherently more accurate than the conventional FD method for the same grid system. The Laplace transform numerical method offers a stable, non-increasing round-off error technique because calculations have to be performed at this time of observation *t* only using a $\Delta t = t$.

3   MODEL VERIFICATION

3.1   *One-dimensional solute transport through layered porous media*

Layered media are a common example of nonhomogeneous media with distinct re-gions. In the first problem, consider a steady unidirectional flow perpendicular to the layers and each layer is considered homogeneous. The movement of a solute in one-dimensional layered porous media with these flow conditions is shown in Figure 1.

   The governing equation of solute movement for each layer is the same as Equation (2) with the value of K_d equal to zero ($R = 1$). It can be written as:

$$\frac{\partial C}{\partial t} = D_x \frac{\partial^2 C}{\partial x^2} - V_x \frac{\partial C}{\partial x}$$

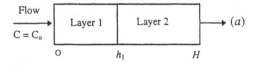

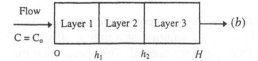

Figure 1. Layered porous media with flow per-pendicular to the stratifications.

subject to the following initial and boundary conditions:

$$C(x, 0) = 0$$
$$C(0, t) = C_0$$

Two-layered and three-layered numerical test problems, shown in Figure 1(a) and Figure 1(b) respectively, were considered in the one-dimensional solute transport problems. The values of H and h_1 are respectively 32 cm and 7.5 cm for both numerical test problems. The value of h_2 equals 15 cm for the problem of the three-layered media numerical test. Three sets of parameter values were used for different cases in each numerical test. Table 1 lists velocities and dispersion coefficients for each layer in each case. $\Delta x = 0.5$ cm was used for the calculation by the Laplace transform numerical method. Figures 2-3 present comparisons of solutions between

Table 1. Velocities and dispersion coefficients for each layer in each case.

	Case					
	1	2	3	4	5	6
V_{x1} (cm /hr)	0.10	0.01	0.05	0.10	0.05	0.10
V_{x2} (cm /hr)	0.05	0.10	0.05	0.05	0.20	0.05
V_{x3} (cm /hr)	–	–	–	0.10	0.10	0.10
D_{x1} (cm^2/hr)	0.18	0.18	0.09	0.18	0.09	0.18
D_{x2} (cm^2/hr)	0.18	0.18	0.90	0.18	0.36	0.08
D_{x3} (cm^2/hr)	–	–	–	0.18	0.18	0.10

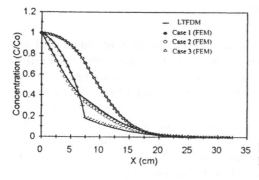

Figure 2. Comparisons of solutions obtained by LTFDM and FEM at time = 100 hours.

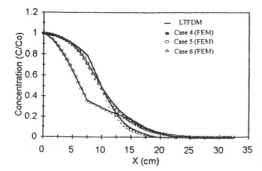

Figure 3. Comparisons of solutions obtained by LTFDM and FEM at time = 100 hours.

the Laplace transform finite difference method (LTFDM) and a Galerkin finite element method (FEM) for each case which show that they are in very good agreement.

3.2 *Two-dimensional solute transport in stratified porous media*

To further verify the Laplace transform numerical method, a two-dimensional solute transport problem (Fig. 4) is considered. Al-Niami & Rushton (1979) derived an analytical solution for dispersion in stratified porous media. Consider a two-layered porous medium in which the direction of flow is parallel to the interface. The difference in groundwater potential between $x = 0$ and $x = L$ (L is the total length of the medium) is the same for each layer, but due to different permeabilities, the velocities in the two layers differ. Effects due to differences in density are ignored. The lateral velocity (V_y) is negligibly small compared with the longitudinal velocity (V_x). However, lateral diffusion is considered to take place across the common boundary between the two layers. The governing equation for this problem in each layer is the same as Equation (3) with $K = 0$ and $V_y = 0$ and subject to the following initial and boundary conditions:

$$C(x, j, 0) = 0 \; ; \; C(0, y, t) = C_0$$

$$\frac{\partial C}{\partial x}(L, y, t) = 0 \; ; \; \frac{\partial C}{\partial y}(x, 0, t) = 0 \; ; \; \frac{\partial C}{\partial y}(x, H, t) = 0$$

On the interface, the condition is:

$$D_{y1} \frac{\partial C}{\partial y}(x, h, t) = D_{y2} \frac{\partial C}{\partial y}(x, h, t)$$

The problem considered was similar to the above one-dimensional numerical test problem, except that the medium has layers extending in the x-direction and a different flow direction. The layers vary in thickness and in medium properties. A porous medium of length 1000 cm and 500 cm total thickness is considered. The same parameter values as presented by Al-Niami & Rushton (1979) are used. The first layer has a thickness (h) of a 125 cm and the fluid flows at a velocity of 0.0004 cm/s. For the second layer, which is 375 cm thick, the fluid velocity is 0.0008 cm/s. The longitudinal dispersion coefficients (D_{xi}) are assumed to be equal and taken as 0.1 cm²/s, while the lateral dispersion coefficient (D_{yi}) in each layer is considered as

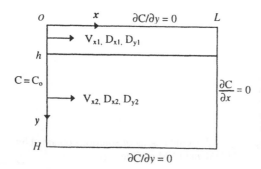

Figure 4. Stratified porous media with flow parallel to the stratifications.

Table 2. Relative concentration distribution for the two-layered stratified porous media at $t = 1523000$ seconds.

y/H	Method	x/L					
		0.0	0.2	0.4	0.6	0.8	1.0
0.00	Analytic	1.0000	0.9272	0.8101	0.6615	0.5178	0.4473
	Wang(1986)	1.0000	0.9296	0.8168	0.6740	0.5367	0.4705
	LTFDM	1.0000	0.9267	0.8112	0.6675	0.5321	0.4673
0.25	Analytic	1.0000	0.9670	0.9090	0.8264	0.7338	0.6808
	Wang(1986)	1.0000	0.9622	0.8966	0.8050	0.7050	0.6500
	LTFDM	1.0000	0.9770	0.9323	.08585	0.7655	0.7057
0.50	Analytic	1.0000	0.9922	0.9707	0.9269	0.8615	0.8151
	Wang(1986)	1.0000	0.9929	0.9727	0.9315	0.8692	0.8244
	LTFDM	1.0000	0.9893	0.9672	0.9208	0.8529	0.8055
0.75	Analytic	1.0000	0.9934	0.9752	0.9382	0.8830	0.8438
	Wang(1986)	1.0000	0.9927	0.9722	0.9305	0.8683	0.8253
	LTFDM	1.0000	0.9893	0.9672	0.9208	0.8529	0.8055
1.00	Analytic	1.0000	0.9936	0.9760	0.9401	0.8866	0.8486
	Wang(1986)	1.0000	0.9927	0.9722	0.9305	0.8683	0.8253
	LTFDM	1.0000	0.9893	0.9672	0.9208	0.8529	0.8055

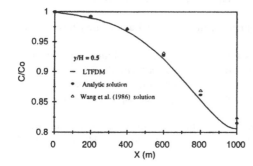

Figure 5. Concentration distribution for the two-layered stratified porous media at $t = 1523000$ seconds.

0.0001 cm^2/s; $\Delta x = 25$ cm, and $\Delta y = 5$ cm were used in the Laplace transform finite difference scheme. The results obtained from the Laplace transform numerical method are compared with the analytical solution of Al-Niami & Rushton (1979) and the numerical solution of Wang et al. (1986) and are shown in Table 2 and Figure 5.

4 CONCLUSIONS

A Laplace transform numerical method for simulating one-dimensional and two-dimensional contaminant movement through stratified porous media has been presented in this paper. The advantage of the Laplace transform numerical method is to eliminate the error caused by the time-discretization treatment of the time derivative in the numerical approximation of the partial differential equation of mass transport problems because the scheme does not involve time steps in Laplace space. This is unlike conventional time-matching schemes where the concentration at time t depends directly on the results computed at $t - \Delta t$. Therefore it is easier to achieve a more accurate result at a desired large time step without excessive execution times. Furthermore, the Stehfest inversion technique involves the use of real numbers only while most of the other techniques require the inversion to be carried out using complex numbers. For this reason, the Stehfest method offers computational advantages over the other methods.

The Laplace transform finite difference method has been shown to produce good agreement with the finite element solutions for modelling the movement of contaminant through layered porous media. It also performed well in predicting the concentration profile for a two-dimensional mass transport problem in stratified porous media. Through comparison with the above one-dimensional numerical tests and a two-dimensional analytical solution, it has been demonstrated that the Laplace transform numerical method is applicable to cases in which the properties of the porous media are stratified.

REFERENCES

Al-Niami, A.N.S. & Rushton, K.R. 1979. Dispersion in stratified porous media: Analytical solutions, *Water Resour. Res.* 15(5): 1044-1048.
Stehfest, H. 1970. Algorithm 368, Numerical inversion of Laplace transforms, *Commun. Assoc. Comput. Mach.*, 13(1): 47-49.
Stehfest, H. 1970. Remark on Algorithm 368, Numerical inversion of Laplace transforms, *Commun. Assoc. Comput. Mach.* 13(10): 624-625.
Sudicky, E.A. 1989. The Laplace transform Galerkin technique: A time-continuous finite element theory and application to mass transport in groundwater, *water resour. res.* 25(8): 1833-1846.
Wang, C., Sun, N.Z. & Yeh, W.W.G. 1986. An upstream weight multiple-cell balance finite-element method for solving three-dimensional convection-dispersion equations, *water resour. res.* 22(11): 1575-1589.
Wang, J.C. & Apperley, L.W. 1994. A Laplace transform numerical model for contaminant transport in porous media, *Proc. 8th Int. Conf. of the Association for Computer Methods and Advances in Geomechanics, Morgantown, West Virginia, USA*: 1177-1182.

CHAPTER 15

Estimation of solute transport parameters in a karstic aquifer using artificial tracer experiments

P. MALOSZEWSKI
GSF, Institute for Hydrology, Oberschleißheim, Germany

R. BENISCHKE, T. HARUM & H. ZOJER
Institute for Hydrogeology and Geothermics, Joanneum Research, Graz, Austria

ABSTRACT: The artifical tracer experiments performed in the Lurbach karst system (north of Graz, Austria) were interpreted using different mathematical models that simulated mass transport between injection and detection sites. The first model assumes that the system consists of several parallel sub-systems characterised by different flow rates, water volumes and dispersivities. The second model assumes that the tracer flows through a double-porosity system consisting of several parallel channels in the microporous matrix, and containing both mobile and quasi-stagnant (immobile) water. The channels have the same apertures, volumetric flow rates and dispersivities and are equally distributed in the matrix. The tracer diffuses between mobile and immobile phase. Both models can be calibrated to fit the experimental data to a high degree of accuracy. Additionally, the measured tracer concentration curves were interpreted by applying the method of moments, a commonly used practice. It is shown herein that the parameters obtained in this way could not make the model fit the experimental data.

1 INTRODUCTION

Knowledge about water storage and flow in karstic groundwater systems plays a particularly important role in countries where drinking water is obtained mainly from karst aquifers. Estimation of exploitable water volume and flow parameters gives significant information for delimiting contamination protection zones. Experimental karst studies include the use of a wide variety of artificial tracers as a tool to determine hydraulic and transport parameters between injection and detection sites. Several well-described tracer experiments performed in karst of different types can be found in, e.g. Behrens et al. (1992), Gospodaric et al. (1976), Morfis & Zojer (1986), and Quinlan et al. (1987). In nearly every case described, the tracer concentration curves are characterised by a strong tailing effect. Only some experiments are interpreted quantitatively and in most cases the method of moments is used, e.g. Stober (1988). Assuming that the tracer transport is described by the one-dimensional dispersion-convection equation, the water velocity (or mean transit time) and dispersiv-

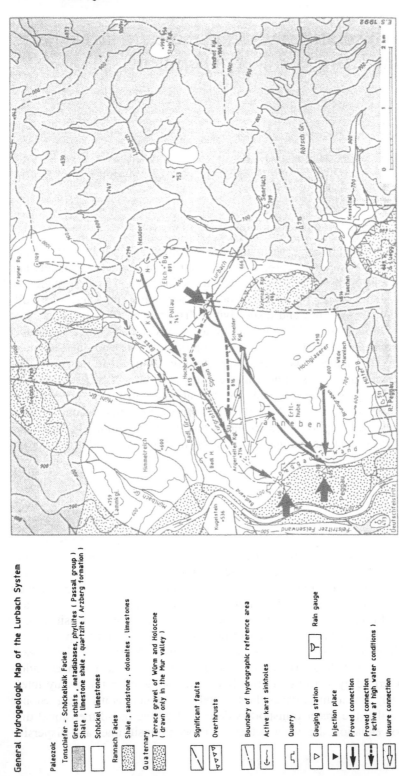

Figure 1. Hydrogeological map of the investigated area – Tanneben karst massif and Lurbach catchment. (HB – Hammerbach spring; S – Schmelzbach spring; the arrows indicate the Lurbach sinkhole and both springs).

ity are calculated directly from the centre of gravity and the time-variance of the tracer concentration curve observed. This two-parameter dispersion model, however, cannot be calibrated when the tracer curve has a very strong tail, which thus renders it non-applicable.

Throughout the cited literature two main modelling approaches are used to describe the strong tailing effects of 'ideal tracers'. The first model assumes that the system consists of several individual, parallel sub-systems (flowpaths) having different water velocities and dispersivities. This approach was developed for porous aquifers by Zuber (1974) and futher developed for karstic systems by Maloszewski et al. (1992). The second model assumes that the tracer flows through a 'double-porosity' system which is approximated by parallel, indentical channels (or fissures) equally distributed in the microporous matrix. The channels include mobile water, whereas stagnant (immobile) water exists in the microporous matrix. The tracer diffuses between these water phases. This approach was introduced for artificial tracer experiments in fissured aquifers by Maloszewski & Zuber (1985) and applied in karstic aquifers by Seiler et al. (1989). The results of applying both models are presented in the following study and are compared with results obtained by applying the method of moments.

2 INVESTIGATED AREA

Situated about 15 km north of Graz (Austria), this study area belongs to the Central Styrian Karst, and has a total area of about 25 km^2 (see Fig. 1). The core of the investigated area includes the catchment of the Lurbach creek (14.5 km^2) and Tanneben karst massif (8.3 km^2). This Devonian limestone block is underlain by schists and phyllites of the same age. The limestone area is bordered by steep faults, whose prevailing joint directions have favoured development of the high, steep rocky face formations that characterize the region's natural scenery. Water disappearing mainly into the Lurbach sinkhole traverses the karst massif of Tanneben and reemerges in the Schmelzbach ($Q = 97$ l/s) and Hammerbach ($Q = 192$ l/s) springs situated 340 m below in the Mur river valley, the bottom of the massif. Normally the Hammerbach spring carries the most discharge, whereas the Schmelzbach runoff fluctuates to a high extent. Until Hammerbach attains a discharge of greater than 200 l/s, both aquifer systems (Schmelzbach and Hammerbach) remain separated and all water from the Lurbach sinkhole reappears soley in the Hammerbach spring. Due to Hammerbach's rather limited storage body, increasing inflow conditions cause it to overflow into the Schmelzbach system.

3 TRACER EXPERIMENTS

The catchment area of both karstic springs is well known due to several tracer experiments performed in different sites of the investigated area (see Fig. 1). Since 1971 twelve tracer experiments have been carried out at the most important sinkhole (Lurbach), primarily utilizing the fluorescent dye tracer Uranine. In the years 1988 and 1993 Uranine was simultaneously injected with Bromide, whereas in 1991 Chlo-

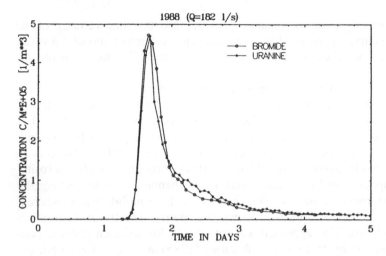

Figure 2. Normalized Bromide and Uranine concentration curves (C/M) observed in Hammerbach spring during a tracer experiment performed in 1988.

ride and Uranine were used. The comparison of Bromide and Uranine tracer concentration curves has shown that Uranine can be considered an ideal tracer for the investigation area (e.g. compare the measured Bromide and Uranine tracer concentration curves shown in Fig. 2). The tracers were injected directly into the Lurbach creek, which then enters the karstic system through the Lurbach sinkhole, and have been observed at both karstic springs (Hammerbach and Schmelzbach) situated approximately 3 km downstream (see arrows in Fig. 1). All tracer experiments from the Lurbach sinkhole were successful in regard to tracer reappearance at the Hammerbach spring. Only during high discharge were the tracers detectable in Schmelzbach spring. All experiments were performed under near-steady state hydraulic conditions (discharge was constant during the tracer flow). The mean volumetric flow rates of water, however, were different in each experiment. Normally, the tracer recovery measured in the Hammerbach spring varied between 60% and 75%, but never reached 100%. In two cases, during very high levels of water discharge when the tracer was detectable in the Schmelzbach spring, the common tracer recovery of both springs approached 100%. This suggests that during low discharges (about 25-40%) injected tracer mass is stored for a longer time in micropores or syphons. All measured tracer concentration curves were characterised by a strong tailing effect. The considerations presented in this study are limited to the Hammerbach system, due to the fact that in most cases the discharges were so low that the tracer concentrations in the Schmelzbach spring were undetectable.

4   MATHEMATICAL MODELS

4.1   *Multidispersion model* (MDM)

The idea of this model is presented in Figure 3. It is assumed that the tracer flows from the system entrance (injection site) to the system exit (detection site) on parallel but different flow paths (flow channels). Each flow path is characterised by a specific volumetric flow rate, water transit time and dispersivity (dispersion parameter). It is

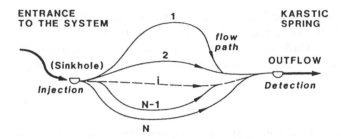

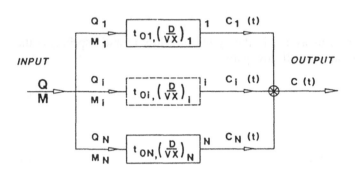

Figure 3. Hydrological (upper) and conceptual (lower) model (MDM) of tracer transport in the Lurbach karstic system.

assumed that there are not interactions between the flow paths, and the possible diffusion of tracer in microporous matrix and/or temporarily nonactive parts of the karstic system are neglected. Under these assumptions the transport of an ideal tracer on the i-th flow path is described by the following one-dimensional dispersion equation:

$$\alpha_i v_i \frac{\partial^2 C_i}{\partial x^2} + v_i \frac{\partial C_i}{\partial x} = \frac{\partial C_i}{\partial t} \tag{1}$$

where $C_i(t)$ is the concentration of tracer in the outflow, α_i is the longitudinal dispersivity, and v_i is the mean water velocity, for the i-th flow path, respectively. The solution to Equation (1) for an instantaneous injection described by the Dirac function has the following form:

$$C_i(t) = \frac{M_i}{Q_i t_{oi} \sqrt{4\pi (D/vx)_i (t/t_{oi})^3}} \cdot \exp\left[-\frac{(1 - t/t_{oi})^2}{4(D/vx)_i (t/t_{oi})} \right] \tag{2}$$

where:

$$(D/vx)_i = \alpha_i / x_i \tag{3}$$

is the dispersion parameter on the i-th flow path, and

$$t_{oi} = x_i / v_i = V_i / Q_i \tag{4}$$

is the mean transit time of water for the i-th flow path, V_i is the mean volume of water and Q_i is the volumetric flow rate, in the i-th flow path, respectively.

The model assumes that the whole mass of tracer injected, M, is divided into N portions which enter the N flow paths proportionally to the volumetric flow rates Q_i. This means that:

$$M_i / Q_i = M / Q \tag{5}$$

where Q is the total discharge of the karstic spring equal simultaneously to the sum of partial flow rates:

$$\sum_{i=1}^{N} Q_i = Q \tag{6}$$

The tracer concentration $C(t)$ measured in the system outflow (karstic spring) is the weighted mean concentration from all flow paths:

$$C(t) = \sum_{i=1}^{N} r_i C_i(t) \tag{7}$$

with

$$r_i = Q_i / Q = M_i / M = R_i / R \tag{8}$$

where R_i is the tracer recovery from the i-th flow path measured in the outflow (karstic spring), whereas R is the total tracer recovery.

Equation (7) combined with Equation (2) is called the Multidispersion Model (MDM) and is used in the fitting procedure to determine all model parameters (solving of the inverse problem). The model's fitting parameters are the mean transit time, dispersion parameter and the portion of water fluxes, for each flow path, and the total number of flow paths. It must be pointed out that, due to the high number of unknowns, the solution of the inverse problem cannot simply be done automatically. It can, however, be done by dividing step by step the experimental curve into the partial curves. The number of flow paths N is then automatically found. After determining the partial volumetric flow rates and transit times of water, the volume of water for each of the N flow paths can be calculated from Equation (4). Their sum is equal to the whole volume of water in such a system:

$$V = \sum_{i=1}^{N} V_i \tag{9}$$

4.2 *Single Fissure Dispersion Model* (SFDM)

Sudicky & Frind (1982) introduced a parallel fissure dispersion model for continuous tracer injection into a fissured groundwater system. The fissured aquifer is approximated by the system of parallel fissures having the same aperture and being equally spaced in the microporous matrix (see Fig. 4). The tracer is injected simultaneously into all fissures, and is transported by convection and dispersion along the fissures. The outflow concentration from all fissures (well mixed) is then measured. The

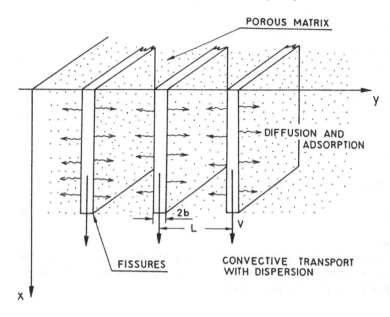

POROUS MATRIX

y

DIFFUSION AND
ADSORPTION

2b

L

v

FISSURES

CONVECTIVE TRANSPORT
WITH DISPERSION

x

Figure 4. Conceptual
model of tracer
transport through a
fissured aquifer ($2b$ –
fissure aperture; L –
fissure spacing; v –
water velocity).

fissures contain mobile water, whereas the microporous matrix contains stagnant (immobile) water. Between the fissures in the matrix the tracer can only be transported by diffusion perpendicular to the flow direction (fissure axis). Maloszewski & Zuber (1985) and Maloszewski (1994) have found that in the case of low water transit times tracer transport in a system of parallel fissures can be described to a good degree of accuracy by the transport in a single fissure situated in the infinitely extended matrix. This approach was used by Seiler et al. (1989) for interpreting the tracer concentration curves of Franken Alb (Germany) karst.

The model is defined by the following equations:

$$\frac{\partial C_f}{\partial t} + v \frac{\partial C_f}{\partial x} - D \frac{\partial^2 C_f}{\partial x^2} - \frac{n_p D_p}{2b} \frac{\partial C_p}{\partial y}\bigg|_{y=b} = 0 \tag{10}$$

for the fissure and

$$\frac{\partial C_p}{\partial t} = D_p \frac{\partial^2 C_p}{\partial y^2} \tag{11}$$

for the microporous matrix.

C_f and C_p are fissure and microporous matrix water tracer concentrations, respectively, v is the mean water velocity in the fissure, D is the dispersion coefficient in the fissures, and x and y are spatial coordinates taken with respect to flow direction and perpendicular to fissure axis, respectively. $2b$, n_p and D_p are the fissure aperture, matrix porosity and diffusion coefficient in water of the microporous matrix, respectively. The solution of Equations (10) and (11) for instantaneous injection of tracer described by the Dirac function and under the assumption that there is no tracer within the system prior to injection, and that fissure spacing L is sufficiently large that there is no interaction between fissures, is given e.g. by Maloszewski (1994) and has the following form:

$$C_f(t) = \frac{aM\sqrt{t_o}}{2\pi Q\sqrt{D/vx}} \int_0^t \exp\left[-\frac{(t_o-\xi)^2}{4D/vx\xi t_o} - \frac{a^2\xi^2}{t-\xi}\right] \cdot \frac{d\xi}{\sqrt{\xi(t-\xi)^2}} \qquad (12)$$

where t_o is the mean transit time of water through the fissures equal to the ratio of volume of water in the fissures (V_f) to the volumetric flow rate of water through the fissures (Q):

$$t_o = V_f / Q \qquad (13)$$

D/vx is the dispersion parameter for the fissures and

$$a = n_p \sqrt{D_p} / (2b) \qquad (14)$$

is the diffusion parameter.

The solution of Equation (12) is called Single Fissure Dispersion Model (SFDM) and has three unknown (fitting) parameters: a, D/vx, and t_o, which must be found through model calibration using experimental data.

4.3 *Method of Moments* (MM)

Throughout the cited literature, e.g. Kreft & Zuber (1978), the method of moments (MM) is well known and is very often used for the interpretation of artificial tracer experiments. From the tracer concentration curve measured as a function of time, $C(t)$, the two following parameters are directly calculated:

The center of gravity (mean transit time of tracer t_t) defined as follows:

$$t_t = \int_0^\infty tC(t)dt \Big/ \int_0^\infty C(t)dt \qquad (15)$$

and the time variance(σ_t^2), defined as:

$$\sigma_t^2 = \int_0^\infty (t-t_t)^2 C(t)dt \Big/ \int_0^\infty C(t)dt \qquad (16)$$

Assuming that the one-dimensional dispersion Equation (2) is valid for the system under consideration it is a common practice to calculate the transport parameters mean transit of water (or water velocity) and dispersion parameter (or dispersivity) from Equations (15) and (16). In such a case the mean transit time of water is directly equal to the mean transit time of an 'ideal' tracer:

$$t_o = t_t \qquad (17)$$

and the dispersion parameter is equal to:

$$D/vx = 0.5\sigma_t^2 / t_t^2 \qquad (18)$$

The application of this method is normally limited to the determination of both t_t and σ_t^2 from the tracer concentration curve measured and to the calculation of t_o and D/vx from Equations (17) and (18).

5 RESULTS OF MODELLING AND DISCUSSION

Twelve experiments performed in the investigation area cover a very wide spectrum of volumetric flow rates through the Hammerbach system. The Hammerbach spring discharge varies between $Q = 80$ l/s and $Q = 360$ l/s and, as mentioned, is nearly constant during each experiment. The MDM and SFDM models were calibrated to fit the tracer concentration curves observed in all twelve experiments with a satisfactory accuracy, whereas the MM does not for any of the experiments. Figure 5 shows an ex-

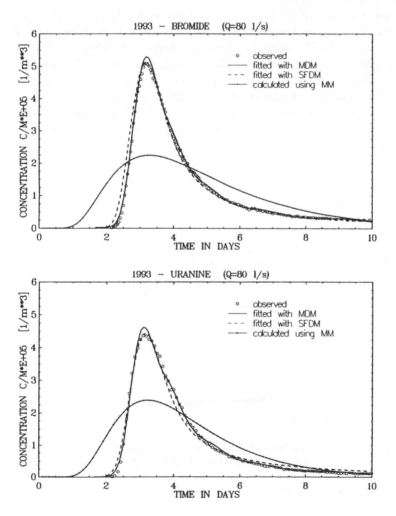

Figure 5. Results of applying different models to Bromide (upper) and Uranine (lower) tracer concentration curves measured in the Hammerbach spring. The model parameters are as follow: Bromide: MDM (t_{oi}: 77, 97, 122, 161, 220hrs, α_i: 24, 18, 20, 22, 25 m; r_i: 0.47, 0.22, 0.15, 0.09, 0.07; for $i = 1,...5$); SFDM ($t_o = 66$hrs, $\alpha = 23$ m, a = 6.7*10^{-4}s$^{-1/2}$); MM ($t_o = 120$ hrs, $\alpha = 475$ m). Uranine: MDM (t_{oi}: 75, 96, 122, 157, 214 hrs; α_i: 24, 18, 20, 20, 30 m; r_i: 0.46, 0.26, 0.15, 0.07, 0.06; for $i = 1,...5$); SFDM ($t_o = 66$ hrs, $\alpha = 26$ m, a = 6.7*10^{-4}s$^{-1/2}$); MM ($t_o = 106$ hrs, $\alpha = 345$ m).

ample of the best fit curves obtained using the MDM and SFDM models for a multi-tracer experiment (simultaneous injection of Bromide and Uranine) performed in 1993 when the mean discharge was $Q = 80$ l/s. The modelled and observed Uranine tracer concentration curves obtained during a relatively high discharge of Hammerbach ($Q = 360$ l/s) are shown in Figure 6.

By applying MDM, up to five independent flow paths were found in the Hammerbach system. The portions of water flux through those flow paths, calculated as an average from all twelve experiments and found to be independent of the volumetric flow rate, are summarized in Table 1.

The volume of water in each of the flow paths, and consequently the system's total water volume calculated by applying Equation (9), was found to be a function of volumetric flow rate. When the Hammerbach spring discharge is roughly $Q = 300$ l/s the total volume of water within the system reaches its maximal value and stays constant for higher discharges. The function $V = f(Q)$ shown in Figure 7 suggests that the capacity of the Hammerbach system is limited to about $V = 40,000$ m^3. Correspondingly, for discharges lower than $Q = 300$ l/s the system is not saturated with water. This is in opposition to the findings of Maloszewski et al. (1992) for the system being considered.

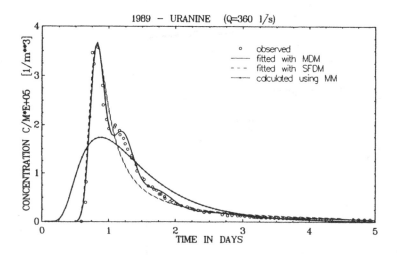

Figure 6. Results of applying different models to Uranine tracer concentration curves measured in the Hammerbach spring. The model parameters are as follow: MDM (t_{oi}: 22, 31, 43, 58, 84 hrs; α_i: 27, 28, 26, 30, 40 m; r_i: 0.44, 0.30, 0.15, 0.07, 0.04; for $i = 1,...5$); SFDM ($t_o = 17$ hrs, $\alpha = 16$ m, a $= 17*10^{-4}$s$^{-1/2}$); MM ($t_o = 34$ hrs, $\alpha = 528$ m).

Table 1. The portions of volumetric flow rate in five flow paths to Hammerbach Spring calculated as an average from the parameters found by applying MDM in 12 tracer experiments.

Flow path i	1	2	3	4	5
Portion of water flux Q through i-th flow path [r_i]	0.52	0.22	0.13	0.07	0.06

By applying the SFDM the volume of water in the system calculated as $V_f = Qt_o$, differs insignificantly from experiment to experiment and seems to be independent of the volumetric flow rate of water (see Fig. 7). These results suggest that the system is permanently saturated with water and has a limited capacity of about $V = 22,600$ m^3.

The mean transit time of water in each of the flow paths and the weighted mean transit time of the whole Hammerbach drainage system calculated as $t_o = V/Q$ is found also to be a function of discharge Q (see Fig. 8) for both MDM and SFDM models.

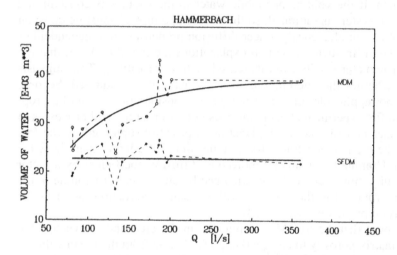

Figure 7. The volume of water in the Hammerbach system as a function of volumetric flow rate of water through the system obtained from applying the MDM and SFDM models.

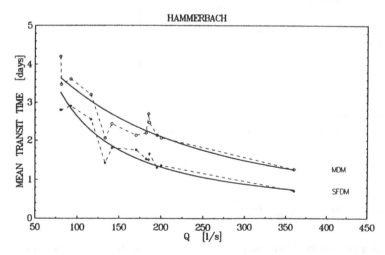

Figure 8. Mean transit time of water through the Hammerbach system as a function of volumetric flow rate through the system obtained by applying the MDM and SFDM models.

It is interesting to point out that if flow path distances of $x = 3300$ m are assumed, the dispersivities are independent of volumetric flow rates for both the MDM and SFDM models. MDM gives dispersivities between 18 m and 26 m, with an average of about 23 m (see Fig. 9). For SFDM dispersivities are also uniform but lower, ranging from 7 m to 17 m with an average of about 13 m.

Of great interest is the fact that the known diffusion parameter a in Equation (14) seems to be a function of volumetric flow rate Q. When the discharge increases the value of the a-parameter also increases (e.g. for Uranine, the a-parameter changes from 5.8×10^{-4} to 17×10^{-4} s$^{-1/2}$ when Q increases from 80 to 360 l/s), a finding that is difficult to explain. If the volume of mobile water in the fissures is constant and independent of Q the system is saturated, so the fissure aperture $2b$ must be constant and independent of Q. Simultaneously, tracer diffusion properties are independent of the hydraulic condition. In such a case two explanations are possible. First, that the increase of the a-parameter results from increased matrix porosity n_p. This suggests that during low discharges the microporous matrix is partially unsaturated. Second, that the SFDM is not applicable for the system under consideration: it can be properly calibrated, and fits experimental data, but does not describe the real processes in the system. The latter half of this last explanation is possibly supported by the fact that the values of the a-parameter found for Uranine were little larger than those for Bromide, although Uranine has a smaller molecular diffusion coefficient by a factor of 3.5. This possibility, however, could be explained by absorption of Uranine in the microporous matrix (but not in the fissures) with the same retardation factor as the difference in diffusion properties (see Eq. 14).

Taking into account that the SFDM is applicable in the area under investigation and assuming the matrix porosity to be $n_p = 0.05$, one obtains from the mean value of diffusion parameter $a = 5.9 \times 10^{-4}$ s$^{-1/2}$ and the known Bromide diffusion coefficient of $D_p = 10^{-9}$ m^2/s, the relatively large value of mean fissure aperture of $2b = 2.8$ mm.

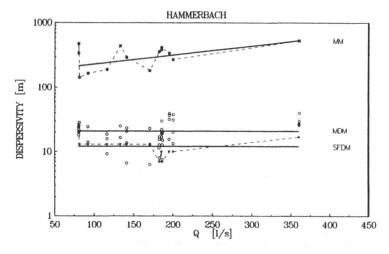

Figure 9. Dispersivity as a function of volumetric flow rate through the system obtained by applying the MDM, SFDM and MM models.

In Figures 5 and 6 are additionally shown theoretical tracer concentration curves that were calculated using a one-dimensional dispersion model with parameters obtained by applying the method of moments (MM). The comparison of those curves with experimental data shows that while the center of gravity and time-variance can be calculated properly, the one-dimensional dispersion model is not able to describe the transport of tracer through the system under consideration. It is interesting, however, to point out that the dispersivities obtained by MM have very high values (10-30 times larger than by MDM) which increase with the discharge (see Fig. 9). This fact is often reported in literature dealing with the interpretation of tracer experiments performed in karstic areas.

6 CONCLUSIONS

The results of modelling show that the MDM and SFDM models are able to be calibrated to fit the experimental data with satisfactory accuracy. On the other hand, a one dimensional dispersion model with only two parameters found by applying the method of moments yields tracer concentration curves that are completely different from the observed concentrations. Simply, a model which cannot be calibrated cannot be applied. This means that the method of moments applied with a one-dimensional dispersion model cannot be used for describing tracer transport in the system under consideration.

Both models, MDM and SFDM, yield nearly the same hydrogeological information; that the water capacity of the Hammerbach system is limited, and that the system has relatively low values of dispersivities that are independent of discharge. The five pathway MDM model better describes the tracer concentration curves. However, the parsimonious 3 parameter SFDM model performs almost as well. From the hydraulic point of view, applying the MDM seems to be more adequate when determining the dependence of water volume within the system on the volumetric flow rate of water through the system (of course, one cannot exclude the possibility that the Hammerbach system is permanently saturated with water as was found from SFDM). However, it should be pointed out that as long as the model cannot be directly validated, it is impossible to decide which of these models better describes the real system.

REFERENCES

Behrens, H., Benischke, R., Bricelj, M., Harum, T., Käss, W., Kosi, G., Leditzky, H., Leibundgut, Ch., Maloszewski, P., Maurin, V., Rajner, V., Rank, D., Reichert, B., Stadler, H., Stichler, W., Trimborn, P., Zojer, H. & Zupan, B. 1992. Investigations with natural and artificial tracers in the karst aquifer of the Lurbach system (Peggau-Tanneben – Semriach, Austria). *Steirische Beiträge zur Hydrogeologie* 43: 9-158.
Gospodaric, R. & Habic, P. (eds). 1976. *Underground Water Tracing – Investigations in Slovenia 1972-1975*. Institute for Karst Research SAZU, Postojna, Yugoslavia.
Kreft, A. & Zuber, A. 1978. On the physical meaning of the dispersion equation and its solution for different initial and boundary conditions. *Chem. Eng. Sci.* 33: 1471-1480.

Maloszewski, P. 1994. Mathematical modelling of tracer experiments in fissured aquifers. *Freiburger Schriften zur Hydrologie* 2: 1-107.

Maloszewski, P., Harum. T. & Benischke, R. 1992. Mathematical modelling of tracer experiments in the karst of Lurbach system. *Steirische Beiträge zur Hydrogeologie* 43: 116-136.

Maloszewski, P. & Zuber, A. 1985. On the theory of tracer experiments in fissured rocks with a porous matrix. *Journal of Hydrology* 79: 333-358.

Morfis, A. & Zojer, H. (eds). 1986. Karst Hydrogeology of the Central and Eastern Peloponnesus (Greece) – 5th International Symposium on Underground Water Tracing, Athens. *Steirische Beiträge zur Hydrogeologie* 37/38: 1-301.

Quinlan, J.F. & Alexander, E.C. 1987. How often should samples be taken at relevant locations for reliable monitoring of pollutants from agricultural, waste disposal, or spill site in karst terrane? First approximation. In: *Proc. of 2nd multidisciplinary conference of sinkholes and the environmental impact of karst*, *Orlando, Florida 1987*, 277-286. Rotterdam: Balkema.

Seiler, K.-P., Maloszewski, P. & Behrens, H. 1989. Hydrodynamic dispersion in karstified limestone and dolomites in Upper Jura of Franconian Alb, FRG. *Journal of Hydrology* 108: 235-247.

Stober, I. 1988. Dispersion als Hinweis auf den Karsttypus. *Deut. Gewässerkundliche Mitt.* 4: 107-110.

Sudicky, E.A. & Frind, E.O. 1982. Contaminant transport in fractured porous media: analytical solution for a single fracture. *Water Resour. Res.* 18: 1634-1642.

Zuber, A. 1974. Theoretical possibilities of the two – well pulse method. In: *Isotope Techniques in Groundwater Hydrology*: 277-294. Vienna: IAEA.

CHAPTER 16

Solute movement in an unsaturated sand under unstable (fingered) flow conditions: Tracer results

C.L. WOJICK & N.J. HUTZLER
Department of Civil and Environmental Engineering, Michigan Technological University, Houghton, USA

J.S. GIERKE
Department of Geological Engineering, Geology and Geophysics, Michigan Technological University, Houghton, USA

ABSTRACT: Unsaturated preferential flow pathways transport contaminants to aquifers at rates faster than predicted by advection-dispersion models. Gravity-driven preferential ('fingered') flow can occur even in relatively homogeneous soils. Bromide and atrazine tracer experiments were conducted within flow fingers induced in a thin-slab box containing uniform sized quartz sand. Duplicate tests were run at selected flow rates and initial moisture contents (dry and drained). Moisture distributions were measured using a light transmission technique. Moisture content analyses of fingers revealed a core of high saturation ($S = 0.4$ to 0.55) within a horizontal moisture content gradient where the moisture content decreased to the initial value. The gradient was much steeper when the sand was initially dry. Solute residence time and flow velocity were determined from measured tracer response data using moment analysis of residence time distributions. The solute transport velocities ranged from 0.19 cm $\sec^{-1}$ for fingers developed in initially air-dry sand to 0.04 cm $\sec^{-1}$ for fingers developed in wetted sand (10% saturation). Tracer response data for flow through individual fingers were compared to advection-dispersion (A-D) model predictions. While the A-D model describes transport through fingers developed in sand that was initially air dry, solute breakthrough for fingers developed in initially wetted sand deviated from advection-dispersion-dominant transport due to the presence of zones of relatively 'stagnant' water.

1 INTRODUCTION

Preferential flow is a term used to describe soil systems where water flows through a fraction of the available wetted area. An obvious cause of preferential transport is the presence of macropores, which result from aggregation, cracks and root holes. However gravity-driven (unstable) preferential flow, or fingered flow, has been observed in unsaturated soils without obvious macropores (e.g. Rice et al., 1986). In general, unstable flow creates conduits for transport that circumvent a significant fraction of the soil thereby moving solutes more rapidly downward than when the entire horizontal area is conducting flow. Figure 1 illustrates a fingered-flow system, which de-

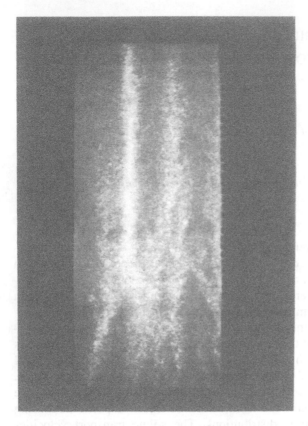

Figure 1. Fingered-flow system developed under conditions of limited infiltration into homogeneous sand. The relative degree of moisture saturation is indicated by image brightness, with higher saturations correlating to the brighter regions. This grey scale image from a multi-finger experiment depicts the steady-state moisture distribution resulting from uniform application of water across the top of initially wetted Ottawa sand (size fraction: 0.60 to 0.85 mm).

veloped under conditions of limited infiltration into homogeneous sand. In this figure, the relative degree of moisture saturation is indicated by image brightness, with highest saturations correlating to the brightest regions.

It is a common practice to assume uniform vertical flow in homogeneous, granular soils, but it is an erroneous assumption for unstable flow conditions. Glass et al. (1989a) and Hillel & Baker (1988) provide a comprehensive history of unstable flow research. Fingered flow can occur under unsaturated conditions in highly-conductive, even sandy, soils. Hillel & Baker (1988) proposed that flow fingers developed when water is supplied to the surface of a highly-conductive soil at a rate less than the conductivity of the soil at its water-entry suction. Their theories have been refined more recently by Liu et al. (1994). Transport processes have different effects when flow is unstable than when it is uniformly distributed.

Non-equilibrium transport processes can result in an apparent faster solute velocity, i.e. early breakthrough, and include such processes as diffusion into immobile zones and rate-limited sorption. Similarly, non-uniform advection rates cause an apparent early breakthrough. Glass et al. (1989b) monitored dye concentrations in unstable flow fields and demonstrated that the processes of advection and dispersion could not consistently account for dye movement.

A protocol was developed for experiments to ascertain the governing processes for solute transport in a gravity-driven flow regime. Previous lysimeter experiments involving solute transport within fingered flow regimes have been limited to single flow fingers or fingers developed in initially dry sand. The experiments described herein include both single- and multiple-finger tests in sands that were initially residually wetted, which is more realistic of field conditions. Steady-state single-finger flow regimes were created to observe the effects of physical non-equilibrium without the influence of multiple fingers. A steady-state multi-finger experiment was conducted to investigate the transport of solutes in individual finger regimes and to compare the finger-specific results to the flux averaged across the lysimeter.

The solute concentration response data were analyzed using moment methods and compared to calculations of an analytic solution for the advection-dispersion model. The analytic solution used herein is more appropriate for the boundary conditions imposed by bench-scale lysimeters then solutions used by previous researchers. Moreover, the analysis of moments described in this paper was more thorough than performed previously and provides comparisons of the mean solute residence time with the hydraulic detention time of the finger system.

2 MATERIALS AND METHODS

Laboratory experiments were conducted to measure the moisture content distribution and tracer transport for gravity-driven preferential flow fingers under different flow rates and initial moisture contents. A schematic of the lysimeter apparatus used for single-finger experiments and the multi-finger experiment is shown in Figure 2. The design of the flow distribution across the top of the lysimeter and the flow collection across the bottom is unique to this work. This design provided uniform flow distribution and the ability to collect solution samples specific to individual finger regimes. In addition, the stones at the bottom of the lysimeter made it possible to establish a relatively uniform initial residual water saturation in the lysimeter. More details of the apparatus design and procedures for the experiments are reported by Wojick (1995).

A thin-section lysimeter constructed of Lucite SAR acrylic (DuPont Co., Willmington, Delaware, USA) with interior dimensions 52-cm high, 23-cm wide, and 1-cm thick, was packed with quartz sand (Ottawa, Illinois, USA). The quartz sand was prepared by acid washing followed by wet sieving to obtain a relatively uniform grain-size ranging between 0.85 and 0.60 mm (US Std. No. 20 × 30 sieves). Special filling procedures were used to produce a uniform packing of the lysimeter with no observed layering of the sand. The porosity range for all experiments was 0.329 to 0.340.

Single flow fingers were established and maintained within the lysimeter by supplying deionized water to a single point at the top of the sand through a small diameter tube (Liu et al., 1994) (single-finger inset in Fig. 2). Constant water application rates, ranging from 1.34 to 3.15 cm^3 min^{-1}, were supplied by a FMI Lab Pump (Fluid Metering, Inc., Oyster Bay, New York, USA). Porous stones installed in the lysimeter base (Inset A in Fig. 2) were used to direct the discharge flow through gravity-suction tubes. Single-finger experiments were conducted in sand that was either initially air dry or residually wetted. The residual moisture content, used to simulate field conditions, was established by saturating the sand with deaireddeionized water and followed by drainage through the porous stones under an applied gravity suction of 50 cm.

Multiple fingers were established and maintained within the lysimeter by supplying deionized water to the fluid distribution chamber mounted on top of the lysimeter (multi-finger inset in Fig. 2). Porous stones installed in this chamber (Inset B in Fig. 2) were used to distribute water uniformly across the sand surface in the lysimeter. Flow rate through the porous stones was controlled by applying a constant head of water, measured in the piezometer tubes, to the chamber. The fluid delivery rate at the applied head of 46 cm was approximately 5.4 cm^3 min^{-1}. Flow at the bottom of the lysimeter was directed through three uniform-size, hydraulically isolated, porous stones installed across the width of the base. The effluent then discharged through the gravity-suction tubes, which were hydraulically connected to each porous stone. The multi-finger experiment was conducted in sand that was initially drained.

Moisture content distributions in the lysimeter were measured using a correlation of backlight transmitted through the lysimeter to moisture content (Glass et al., 1989a; Hoa, 1981). A CCD video camera was used in combination with frame-grabber software (Decision Images, Inc., Skillman, New Jersey, USA) to record light transmission through the lysimeter from a bank of fluorescent bulbs (Fig. 2). An increase in transmitted light is associated with a higher degree of water saturation. Background correction for light source and sand packing variability was achieved by subtracting the recorded air-dry sand image. Image processing software (PV Wave P&C, Precision Visuals, Inc., Boulder Colorado, USA) and graphical software (XV 3.00, John Bradley, Bryn Mawr, Pennsylvania, USA) were used for image calibration and processing for moisture content analyses.

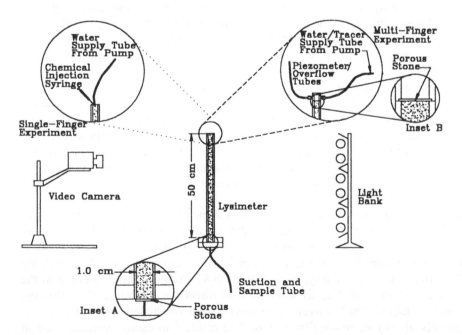

Figure 2. Schematic of laboratory apparatus used for single- and multi-finger experiments.

The primary shortcoming of this technique is the difficulty in calibrating measured transmitted light to water content (Hoa, 1981). A method of correlating measured light intensity to moisture content in thin-section lysimeters containing sand was described by Tidwell & Glass (1994). This method utilizes empirical data in combination with the theoretical relation:

$$S = \frac{\ln[I_{vn}[(\tau_{sw}/\tau_{sa})^{2k} - 1] + 1]}{2k \ln[\tau_{sw}/\tau_{sa}]} \tag{1}$$

where S is the water saturation, τ_{sw} and τ_{sa} are the fractions of light transmitted through the sand-water and sand-air interfaces, k is an empirically fitted parameter determined using a lysimeter image and the gravimetrically determined water content corresponding to this image, and I_{vn} is the normalized net light intensity.

For selected experiments, the lysimeter was weighed prior to the start of finger flow and immediately after termination of the experiment. Gravimetric water contents were used to evaluate the accuracy of moisture contents determined using the light transmission technique and to determine the hydraulic detention time of the system.

2.1 *Single-finger experiments*

For single-finger experiments, mean solute residence times and linear flow velocities were evaluated using measured response data from the injection of solute slugs into the finger influent stream at steady-state flow conditions. Potassium bromide tracer was introduced in concentrated solution (2500 to 6400 mg Br^{-} L^{-1}) over a 20 sec interval. The injected volume (0.1 or 0.05 cm^3) increased the flow rate by less than 10%. In other experiments, atrazine (Crescent Chemical Co., Hauppauge, New York, USA), a common herbicide, was introduced at a concentration of 31 mg L^{-1} under similar injection interval and volume conditions. The injections of concentrated tracer were performed to simulate an instantaneous input or slug.

Effluent samples were collected in 1.0 cm^3 volumes from the suction tube at regular time intervals. For bromide analysis, 1.0 cm^3 of ionic strength adjuster (1 M NaNO$_3$) was added to each sample before analyzing for Br^{-} concentration with an ion selective electrode meter (Orion Research, Inc., Boston, Massachusetts, USA). Atrazine concentrations were measured using direct aqueous injection of effluent samples into a Hewlett-Packard 5890 Gas Chromatograph equipped with a Hewlett-Packard 5970 Mass Selective Detector (Hewlett-Packard Co., Naperville, Illinois, USA) and J&W Scientific DB-FFPA 30 m capillary column (J&W Scientific, Folson, California, USA). Sample collection times were adjusted downward to account for the retention time in the apparatus dead volume (tubing, fittings, and porous stone volumes) and slug injection and effluent sampling time periods.

Measured response concentration versus time data (C, t) were analyzed using moment and area methods (Fahim & Wakao, 1982; Levenspiel, 1979) to obtain values for residence time of solute in the finger (τ), solute velocity (v), mass of bromide collected (M), and the time that halves the area under the response data curve $(t_{1/2})$ (Turner, 1972). Trapezoidal-rule and composite-area methods were used to integrate the area and calculate the moments. The moment calculations correspond to an instantaneous injection, but the actual tracer application was a 20 sec pulse. However, the differences between the slug and pulse injections were found to be negligible for these experiments.

For slug inputs, the residence time distribution or exit age distribution ($E(t)$) represents the average time a certain mass fraction of solute spends in the apparatus:

$$E(t) = \frac{QC(t)}{M_0} \tag{2}$$

where Q is the volumetric flow rate and M_0 is the mass of solute in the injected slug.

Mass conservation during the experiment was evaluated using:

$$M = \int_0^\infty QC(t)dt \tag{3}$$

which is equal to the zero moment (Levenspiel, 1979).

The mean solute residence time (τ) for an instantaneous slug input is equal to the first moment of the residence time distribution (Fahim & Wakao, 1982):

$$\tau = \int_0^\infty E(t)t\,dt \tag{4}$$

In subsequent analyses reported here, solute transport is assumed to be one-dimensional (vertical) and, therefore, the average solute velocity (v) is equal to the travel distance (L) divided by the mean residence time (τ).

2.2   Multi-finger experiment

For the multi-finger experiment, flow parameters were evaluated from solute concentrations measured in effluent samples collected from the suction tubes. This effluent response was measured following the introduction of a step input of tracer to the fluid distribution chamber mounted on top of the lysimeter. The tracer solution for the breakthrough experiment was introduced by rapidly exchanging the water in the fluid distribution chamber with the tracer solution at a pump flow rate of 120 cm^3 min^{-1}. The exchanged fluid was discharged from the chamber via two 0.64 cm piezometer tubes (Fig. 2b). After approximately 2 minutes at the high flow rate, the pumping rate was decreased until the head elevation returned to the desired level of 46 cm. The tracer solution contained potassium bromide at a concentration of approximately 6800 mg Br$^-$ L^{-1}. Elution of the solute was carried out by introducing deionized water following the procedure described above for breakthrough. The step tracer input was used in the multi-finger experiment for ease of application and to provide additional insights into finger-flow patterns by comparison with results from the slug type input. Previous studies have only performed slug inputs.

Effluent samples, collected over 30 sec time intervals, were simultaneously obtained from each of the three sample tubes. Samples were prepared for analysis by first diluting with 3.5 cm^3 of deionized water and then adding 3.9 cm^3 of ionic strength adjuster (1 M NaNO$_3$). Bromide concentrations were measured using an ion selective electrode in the manner described for single-finger experiments. Sample collection times were adjusted to account for the retention time in the apparatus dead volume (tubing, fittings, and porous stone volumes) and effluent sampling time periods.

Measured response data (C, t) were analyzed to obtain values for residence time of solute in the fingers (τ), mass of bromide collected (M), and solute velocity (v). The trapezoidal rule method was used to integrate areas. Assumptions for the analyses include: 1. A conservative tracer, 2. $Q_{in} = Q_{out}$, and 3. Solute transport in the entrance and exit streams of the vessel by advection only (closed vessel conditions, Levenspiel, 1979).

To check mass conservation, the mass of solute in the effluent (M), determined using Equation 3, was compared to the mass of solute introduced before the change to elution (M_0). The mean residence time of the fluid (τ) during breakthrough was calculated from best-fit curves applied to multi-finger elution data using (Levenspiel, 1979):

$$\tau = \frac{\int_0^\infty (C_{max} - C(t))dt}{C_{max} - C_{min}} \tag{5}$$

where C_{max} is the maximum and C_{min} is the minimum concentration observed.

The mean residence time of the fluid during elution was calculated using a variation of Equation 5. The average solute velocity (v) was determined using the procedure described for single-finger experiments.

3   RESULTS AND DISCUSSION

3.1   *Single-finger experiments*

Six single-finger experiments were performed. Experiments 1, 2 and 3 were conducted using a bromide tracer in three primary steps:
1. Establishment of finger flow,
2. Introduction of the first tracer slug A, and
3. After elution of slug A introduction of slug B.

Experiments 1 and 2 were performed with the sand initially air dry. Experiment 3 was conducted with the sand initially drained ($S_0 = 0.11$). Experiments 4 and 5 were conducted with one slug of atrazine tracer used in each experiment. Experiment 4 was performed with the sand initially air dry. For Experiment 5, the sand was initially drained. Experiment 6 was run without tracer in drained sand to obtain data for the moisture content analysis.

The moisture content distribution for Experiment 4, as determined by the light transmission technique, is shown in Figure 3. The distributions for the remaining experiments conducted in air-dry sand (1A, 1B, 2A and 2B) were very similar to that shown in Figure 3.

Figure 4a depicts the moisture distribution for Experiment 3B, which was similar to that for 3A. The moisture content distribution for Experiment 5 is shown in Figure 4b. Even though the hydraulic conditions for Experiments 3 and 5 were nearly the

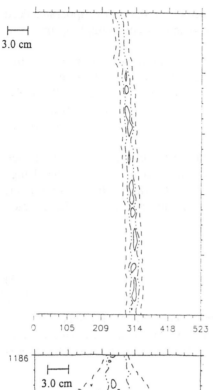

Figure 3. Example of steady-state moisture distribution during infiltration into initially air-dry 20 × 30 Ottawa sand. The dashed contour is the limit of wetting. The solid and dash-dot contours correspond to saturations of 0.45 and 0.30, respectively.

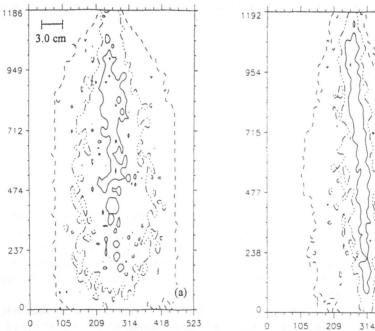

Figure 4. Steady-state moisture distributions at infiltration rates of 3 cm^3 min^{-1} into initially drained ($S_0 = 0.1$) 20 × 30 Ottawa sand for (a) Experiment 3B and (b) Experiment 5. The solid, dash-dot and dashed contours correspond to saturations of 0.40, 0.30, and 0.20, respectively. In (a) the vertical portions of the dashed lines correspond to the periphery of the recorded image rather than the 0.20 saturation contour.

same, moisture distributions for the separate infiltration experiments in residually wetted sand varied markedly. Drainage periods for the residually wetted sand in the lysimeter were 40 and 285 hours for Experiments 3 and 5, respectively. Although the difference in saturations measured for the residually wetted sand was minor (Table 1), additional experiments confirmed that a narrower finger developed for conditions of extended drainage.

Bromide and atrazine concentrations were converted to residence time using Equation 2 and are plotted in Figure 5a and 5b for the initially dry and wet experiments, respectively. Effluent samples were not collected during the initial breakthrough of atrazine in Experiment 5, because chemical breakthrough occurred much earlier than expected based on prior experiments in drained sand. Experiments in air-dry sand and replicate Experiments 3A and B were reproducible. Glass et al. (1989b) reported similar findings.

Flow, water content, and tracer results are reported in Table 1. Mass recovery imbalances of solute for the experiments ranged from -6.9% to $+1.4\%$. Differences between the gravimetrically determined mass of water in the lysimeter and the mass estimated using the light transmission procedure were 21.3%, 15.9%, and 10.8% for Experiments 3B, 4, and 5, respectively.

Table 2 summarizes the results of moment and area analyses. The calculated mean residence times demonstrate the reproducibility of replicate experiments. Average solute velocities (v) were estimated from the finger length (L) and observed residence time (τ). For a finger developed in air-dry sand, the mean residence time and hydraulic detention time are approximately equal (Experiment 4). This indicates that the tracer solute passes through nearly all of the finger region occupied by water.

Table 1. Flow, water content and tracer results for single-finger experiments in 20×30 Ottawa sand. Temperatures for the experiments were $23 \pm 1°C$.

Exp. No.	Flow[1] Q (cm^3 min^{-1})	Initial saturation[2] S_0	Mass tracer injected[3] M_0 (mg)	Mass tracer recovered[4] M (mg)	Mass water in lysimeter[5] W (g)	Estimated water in lysimeter[6] W_e (g)
1A	3.06	0	0.283	0.273	–	–
1B	3.15	–	0.217	0.220	–	–
2A	1.34	0	0.321	0.229	–	–
2B	1.39	–	0.323	0.304	–	–
3A	2.92	0.11	0.682	0.656	–	–
3B	2.96	–	0.731	0.717	119.0	93.6
4	3.05	0	0.00358	0.00353	13.8	11.6
5	2.89	0.11[2]	0.00350	0.00174 [7]	86.9	77.5
6	2.95	0.10	–	–	72.0	Fit to 72.0 [8]

[1]measured directly from suction tube; [2]determined gravimetrically at the beginning of each experiment or set of experiments. Saturation for Experiment 5 was measured after 48 hours of drainage rather than immediately before finger flow as in Experiments 3 and 6. Drainage periods were approximately 40, 285, and 221 hours for Experiments 3, 5, and 6, respectively; [3]determined from concentration and volume of injection slug; [4]from Equation (3); [5]measured gravimetrically at the end of the experiment; [6]determined from integration of the moisture content distribution, which was estimated using Equation (1); [7]effluent samples were not collected during the initial solute breakthrough; [8]measured mass of water used to calibrate k in Equation (1).

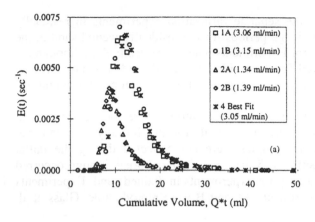

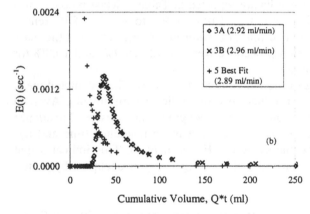

Figure 5. Residence time distributions (E) of bromide and atrazine for experiments in: (a) initially dry and (b) initially drained ($S_0 = 0.1$) 20 × 30 Ottawa sand. Data plotted for Experiments 4 and 5 were best fit to replicate measurements of effluent samples.

Table 2. Results of moment and area analyses for single-finger experiments in 20 × 30 Ottawa sand. (Note: $L = 51.0$ to 51.3 cm for listed experiments).

Exp. no.	Mean residence time[1] τ (min)	Hydraulic detention time[2] τ_g (min)	Time at 1/2 of zero moment[3] $t_{1/2}$ (min)	Average solute velocity[4] v (cm sec^{-1})
1A	4.42	4.19[5]	4.05	0.193
1B	4.48	–	4.15	0.191
2A	8.14	–	7.19	0.105
2B	8.06	–	7.10	0.106
3A	22.1	–	17.0	0.039
3B	22.6	> 40.3[6]	16.9	0.038
4	4.64	4.54	4.34	0.184
5	–	> 30.1[6]	5.7[7]	–

[1]from Equation (4); [2]determined as mass of water added to lysimeter during fingered flow experiment divided by flow rate (Q); [3]determined as 1/2 of area under C versus t curve; [4] = L/τ; [5]calculated using measured value of $t_{1/2}$ and fitted $Pe = 27$; [6]reported value is minimum detention time as the calculation does not account for mobilization of or diffusion in the residual water present in the lysimeter before finger formation; [7]estimated using measured mass of atrazine injected and collected.

This contrasts with experiments conducted in dr ained sand where the mean solute residence time is much less than the hydraulic detention time (Experiment 3B). In this case, the solute slug passes through only a portion of the finger-wetted region.

The analytical solution for the advection-dispersion equation with Dirac δ (slug) input and closed vessel boundary conditions (Thomas & McKee, 1944) was used to model the concentration data from the bromide and atrazine experiments:

$$E(\theta) = 2 \sum_{n=1}^{\infty} \frac{\bar{a}_n \exp\left\{-\left(\frac{\bar{a}_n^2}{Pe} + \frac{Pe}{4}\right)\theta + \frac{Pe}{2}\right\}}{\bar{a}_n^2 + \frac{Pe^2}{4} + Pe} \left(\bar{a}_n \cos\bar{a}_n + \frac{Pe}{2}\sin\bar{a}_n\right) \qquad (6)$$

where Pe is the Peclet number ($v \cdot L/D$) and $\bar{a}_n$ the positive non-zero roots of:

$$\tan\bar{a}_n = \frac{Pe\,\bar{a}_n}{\bar{a}_n^2 - \frac{Pe^2}{4}} \qquad (7)$$

Closed vessel boundary conditions were assumed for the advection-dispersion model due to the relatively small dead volume of the apparatus (3.5 cm^3) compared to the volume of a flow finger (> 100 cm^3). For mathematical convenience, previous studies of solute transport with fingered flow have applied the solution of the advection-dispersion model for open vessel boundary conditions. Although the differences in the two approaches are negligible for high Pe numbers (> 40), the observed values of Pe in this and previous work are low enough to result in significant differences between the solutions for closed and open vessel conditions.

Model calculations for Equation (6) corresponding to $Pe = 27$ are plotted in Figure 6 along with effluent measurements from Experiment 1A and 4, which were conducted in initially air-dry sand, and 3B in initially drained sand. The selected Peclet value of 27 was calibrated by matching the peak concentration predicted by Equation (6) to the maximum concentration observed in Experiment 1A. For the experiments plotted in Figure 6, the time (t) axis was normalized by the hydraulic detention time

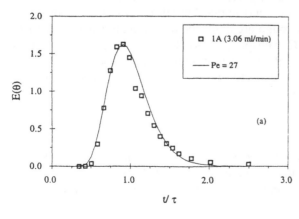

Figure 6. Comparison of advection-dispersion model to observed solute concentrations: (a) Exp. 1A, initially dry.

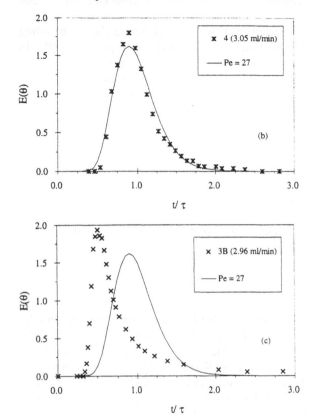

Figure 6. Continued. (b) Exp. 4, initially dry; and (c) Exp. 3B, initially drained.

(τ_g) for experiments in dry sand and the mean solute residence time (τ) for the initially drained experiment. Because errors in the tail portions of breakthrough curves are magnified in first moment calculations, gravimetrically determined moisture content data provides a more accurate measurement for the residence time of a finger developed in dry sand.

The advection-dispersion model closely predicts solute transport for fingers developed in initially air-dry sand (Figs 6a and b). A small amount of elution tailing beyond that predicted from Equation (6) is observed in both of these plots. For fingers developed in initially drained sand, solute breakthrough is advanced from the advection-dispersion model prediction, and noticeable tailing is evident. Long elution tailing and early solute breakthrough are evidence of relatively stagnant regions of water within the finger system (Levenspiel, 1979). This is substantiated by comparison of the mean residence and hydraulic detention times for Experiment 3B (Table 2). Tailing may also result from diffusion of solute between the flow finger and peripheral regions (Johnstone, 1995).

3.2 *Multi-finger experiment*

One multi-finger experiment (Experiment 7) was conducted in initially drained sand. Bromide tracer was introduced to the developed system of fingers in the form of a step input and then eluted. Effluent samples were recovered through the three segregated porous stones in the lysimeter base. A grey scale image, developed from light transmission data, depicting the moisture distribution is shown in Figure 1. The relative degree of moisture saturation is indicated by image brightness, with highest saturations correlating to the brightest regions. Saturations of 0.45 with small zones to 0.60 were measured within the finger shown just right of center in Figure 1. Background saturation of 0.25 to 0.30 were measured across the darker regions. Spacing of the developed fingers was relatively uniform at the bottom of the lysimeter. Finger spacings of 5.1 to 5.8 cm were measured. This resulted in the primary discharge of flow from each finger through a single collection port.

Breakthrough and elution data for each of the sample ports are plotted in Figure 7. A weighted concentration curve is also shown. This is the concentration that would be obtained at a sample time by mixing the effluent from the individual ports. Average flow rates for each sample port are listed in the figure legend. Solute breakthrough and elution occurred more rapidly in ports receiving larger flow rates. Solute mass recovery was determined using Equation (3). The calculated recovery imbalance percentage was -1.7% ($M = 9644.6$ mg, $M_0 = 9810.0$ mg). Breakthrough data plotted on a cumulative volume basis are shown in Figure 8. When analyzed on a

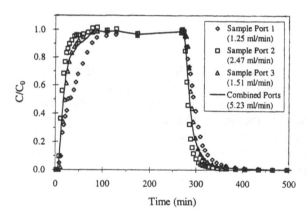

Figure 7. Solute breakthrough and elution data for the multi-finger experiment in initially wetted ($S_0 = 0.12$) 20×30 Ottawa sand. Flow rate data are averages for breakthrough and elution.

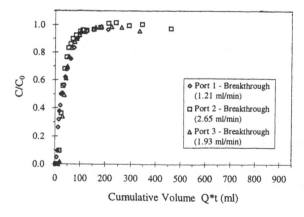

Figure 8. Breakthrough data for the multi-finger experiment plotted on the basis of cumulative flow volume.

Table 3. Flow rate and tracer data for the multi-finger experiment in 20 × 30 Ottawa sand. (Note: L = 51.7; S_0 = 0.12).

Port number[1]	Average flow rate[2] Q (cm^3 min^{-1})	Mean residence time[3] τ (min)	Average solute velocity[4] v (cm sec^{-1})
1-Breakthrough	1.21	38.5	0.022
2	2.65	18.0	0.048
3	1.93	25.4	0.034
Combined	5.79	24.0	0.036
1-Elution	1.28	36.9	0.023
2	2.44	20.4	0.042
3	1.40	31.0	0.028
Combined	5.12	27.2	0.032

[1]referenced from the left – 1, 2, and 3; [2]determined using gravimetric measurements of suction tube samples; [3]from Equation (5); [4] = L/τ.

cumulative flow basis, breakthrough for the individual ports occurs at a consistent rate. Similar results were observed for the elution data. Flow and tracer data for Experiment 7 are listed in Table 3.

4 CONCLUSIONS

Tracer tests in single-finger flow experiments in a uniform sized sand were reproducible. Solute transport through fingers developed in initially drained sand was affected by relatively stagnant zones of water within the finger and the diffusive exchange of solute with areas peripheral to the finger. The advection-dispersion model, while closely predicting single-finger transport in dry sand, could not simulate effluent concentrations for fingers developed in initially drained sand. For the multi-finger experiment, step-input response data for the various flow fingers, when normalized using individual finger-region flow rates, follow similar curves for breakthrough and elution. The light transmission technique is useful for estimating moisture contents in thin-section lysimeters.

ACKNOWLEDGMENTS

This material is based on work supported by the Cooperative State Research Service, US Department of Agriculture, under Agreement No. 92-34214-7385.

REFERENCES

Fahim, M.A. & Wakao, N. 1982. Review paper: parameter estimation from tracer response measurement. *Chem. Eng. J.* 25(1): 1-8.

Glass, R.J., Steenhuis, T.S. & Parlange, J.-Y. 1989a. Mechanisms of finger persistence in homogenous, unsaturated media: Theory and verification. *Soil Sci.* 148(1): 60-70.

Glass, R.J., Oosting, G.H. & Steenhuis, T.S. 1989b. Preferential solute transport in layered homogeneous sands as a consequence of wetting front instability. *J. Hydrol.* 110: 87-105.

Hillel, D. & Baker, R.S. 1988. A descriptive theory of fingering during infiltration into layered soils. *Soil Sci.* 146(1): 51-56.

Hoa, N.T. 1981. A new method allowing the measurement of rapid variations of water content in sandy porous media. *Water Resour. Res.* 17(1): 41-48.

Johnstone, T.L. 1995. Modeling solute transport under fingered flow conditions: A two-dimensional, two-domain approach. M.S. Thesis, Department of Civil and Environmental Engineering, Michigan Technological University, Houghton, Michigan, USA.

Levenspiel, O. 1979. *The Chemical Reactor Omnibook.* Oregon State University, Corvallis, Oregon, USA.

Liu, Y., Steenhuis, T.S. & Parlange, J.-Y. 1994. Closed-form solution for finger width in sandy soil at different water contents. *Water Resour. Res.* 30(4): 949-952.

Rice, R.C., Bowman, R.S. & Jaynes, D.B. 1986. Percolation of water below an irrigated field. *Soil Sci. Soc. Amer. J.* 50(4): 855-859.

Thomas, Jr., H.A. & McKee, J.E. 1994. Longitudinal mixing in aeration tanks. *Sewage Works Journal* 16(1): 42-55.

Tidwell, V.C. & Glass, R.J. 1994. X-ray and visible light transmission for laboratory measurement of two-dimensional saturation fields in thin-slab systems. *Water Resour. Res.* 30(11): 2873-2882.

Turner, A.G. 1972. *Heat and Concentration Waves.* Academic Press, New York, USA.

Wojick, C.L. 1995. *Preferential transport of solutes through unsaturated sand: laboratory experiments.* M.S. Thesis, Departme it of Civil and Environmental Engineering, Michigan Technological University, Houghton, Michigan, USA.

CHAPTER 17

Using HST3D to model flow and tracer transport in an unconfined sandy aquifer with large drawdowns

D. LASLETT & G.B. DAVIS
CSIRO Land and Water, Private Bag, Wembley, Australia

ABSTRACT: Three dimensional groundwater flow and solute transport simulation of an unconfined sandy aquifer with an intensive water pumping and injection regime, and several applied tracer tests, was carried out using the finite difference computer code HST3D. To reduce storage requirements and execution times, and to enable manipulation and visualisation of large output data sets, new matrix solvers and dedicated pre- and post-processor software were developed. A limitation of HST3D is that unconfined aquifers with large drawdowns cannot be modelled without losing fine-scale information near the water table. To overcome this, a double zone confined aquifer with a high storage coefficient in the upper layer was used to mimic the behaviour of the unconfined aquifer. Its behaviour was compared to that of the unconfined aquifer and found to provide an acceptable quantitative match in head and velocity values. The simulation was calibrated by comparison with water levels and bromide tracer test data from a field site. However, it was found that no one set of hydrological parameters could adequately represent conditions over the whole simulation region.

1 INTRODUCTION

The computer program HST3D (Heat and Solute Transport in 3 Dimensions) uses finite difference approximation techniques to solve transient equations describing groundwater flow and solute transport in a fully saturated, three dimensional domain. The code is large (~12,000 lines) and was developed for mainframe use by Kipp (1987). In this paper, HST3D has been modified and applied to a small-scale tracer test site where large changes in water tables were induced in a shallow unconfined aquifer underlying the site. Aquifer parameters were estimated from calibration of the model to field data, which aided design of remediation strategies for enhanced biodegradation of non-aqueous phase liquid (NAPL) diesel that had leaked and contaminated the site (Johnston & Patterson, 1994 and Davis et al., 1993).

HST3D has been used successfully in the past to model groundwater flow and solute transport in unconfined aquifers (Davis & Salama, 1995), where water table changes have been small over the modelled region. However, where water table changes are large, fine-scale information in the near water table zone is lost, since HST3D and other saturated flow codes require that the water table surface always be contained within the top layer of the model grid, and therefore the top layer grid can be excessively thick. For the field tracer test described here, water table changes are large. To allow application of HST3D, dual zone confined aquifer simulations were designed to mimic the behaviour of the unconfined aquifer simulations and thus allow fine-scale vertical grids. Groundwater heads and tracer test data were used to calibrate the confined aquifer simulations and aquifer properties were determined from these simulations. In this paper we also describe modifications to HST3D and specially developed pre- and post- processors that enable large model runs to be efficiently computed.

2 MATRIX SOLVERS

Finite difference techniques enable flow and transport equations to be approximated by linear simultaneous equations. For N nodes, a matrix of N simultaneous equations of N variables must be solved for each time step for both the flow and transport equations. In its original form HST3D provided two matrix solvers: An alternating diagonal direct equation solver (D4) and an iterative successive overrelaxation solver (L2SOR) (Kipp, 1987). For the number of nodes used in the simulations (N ~ 20K to 100K), these solvers proved to be too slow and required too much computer memory storage for practical use.

Two new iterative solvers, based on more recently published matrix solution techniques, were developed and installed in the HST3D computer code. The first solver uses a diagonally preconditioned conjugate gradient algorithm (DCG) and the second solver uses the Seidel (or Jacobi) iterative algorithm (Buckingham, 1962). It was found that the DCG solver gave best results for solving the flow equation matrix, and the SEIDEL solver for the solute equation matrix.

Benchmark testing indicated that the new solvers gave an improvement of 4:1 in solution speed and 3.6:1 in memory storage reduction over the D4 method at N = 4200 nodes. For larger simulations, where memory size is a major limiting factor, the use of the new solvers gave an overall improvement of greater than 50:1. A short description of the matrix solution techniques follows.

2.1 *The diagonally preconditioned conjugate gradient* (DCG) *algorithm*

When calculating changes in water pressure from the flow equation, the HST3D program is required to solve a matrix equation of the following form:

$$A \cdot x = b \tag{1}$$

If the number of nodes in the simulation is N, then A is an $N \times N$ matrix, b is a constant vector of length N, and x is the solution vector of pressure changes of length N. For any matrix A, Equation (1) will be satisfied when:

$$F(x) = \frac{1}{2}|A \cdot x - b|^2 = 0 \tag{2}$$

or

$$\nabla F(x) = A^T \cdot (A \cdot x - b) = 0 \quad (A^T \text{ is the transpose of } A) \tag{3}$$

If A is symmetric and positive definite, then the following function can be used in place of $F(x)$:

$$f(x) = c - b \cdot x + \frac{1}{2} x^T \cdot A \cdot x \tag{4}$$

where c is a constant (Press et al., 1987). $f(x)$ is minimized when:

$$\nabla f(x) = A \cdot x - b = 0, \text{ or } A \cdot x = b \tag{5}$$

The conjugate gradient matrix solution method attempts to minimize $f(x)$. From a starting vector x^0, an iterative loop is used to converge on an x^n vector that satisfies:

$$|b - A \cdot x^n| < e \tag{6}$$

where e is a specified tolerance.

The unpreconditioned conjugate gradient method may converge slowly if the A matrix is ill-conditioned (Meyer et al., 1989), i.e. if the ratio of the largest and smallest eigenvalues of the matrix (the condition number) is large (Press et al., 1987). The condition number of A can be reduced by multiplying by an appropriate preconditioning matrix C, often chosen to be D^{-1} (Meyer et al., 1989), where D is the diagonal matrix of A:

$$D_{ii} = A_{ii}, \ D_{ij} = 0 \ \ i \neq j$$

$$C_{ii} = D_{ii}^{-1} = \frac{1}{A_{ii}}, \ \ C_{ij} = 0 \ \ i \neq j \tag{7}$$

The iteration equations for the diagonally preconditioned conjugate gradient method (DCG) can now be assembled:

$$x^0 = D^{-1} \cdot b, \ g^0 = b - A \cdot x^0, \ h^0 = D^{-1} \cdot g^0$$

$$1^i = \frac{D^{-1} \cdot g^i \cdot g^i}{h^i \cdot A \cdot h^i}, \ x^{i+1} = x^i + 1^i h^i$$

$$g^{i+1} = g^i - 1^i A \cdot h^i \tag{8}$$

$$g^i = \frac{g^{i+1} \cdot g^{i+1}}{g^i \cdot g^i}, \ h^{i+1} = D^{-1} \cdot g^{i+1} + g^i h^i$$

where $g^i = -\nabla f(x^i)$, h^i is the next direction vector of movement from x^i, and 1^i and g^i are constants of proportionality. A full theoretical development of Equation (8) can be found in Laslett & Davis (1993). Note that $f(x)$ is never explicitly computed and that only one matrix multiplication $(A \cdot h^i)$ is performed in each iteration. The iteration loop stops when Equation (6) is satisfied, or when $g^i \cdot g^i < e^2$.

The DCG algorithm developed for HST3D converged much faster than a previously developed unpreconditioned conjugate gradient algorithm. For the water flow equation, the convergence tolerance, e, was set at 1×10^{-8}. Within HST3D, an N node simulation region generates a sparse A matrix requiring storage for 7N elements (instead of N^2). Because the D^{-1} matrix is the inverse of the A matrix diagonal (Eq. 7), the DCG method requires storage for only 2N extra elements (the g^i and h^i arrays).

2.2 *The* SEIDEL *algorithm*

When calculating changes in mass fraction from the solute transport equation, HST3D solves an equation similar to Equation (1), but with an antisymmetric A matrix (non-zero dispersion coefficients might cause A to become asymmetric rather than antisymmetric). The DCG algorithm converged slowly in this case. The simple SEIDEL iterative algorithm (Buckingham, 1962), was found to converge quickly. If D is a matrix containing the diagonal elements of A, and E is a matrix containing the off diagonal elements of A $(= D + E)$:

$$D_{ii} = A_{ii} \quad D_{ij} = 0 \quad i \ne j$$

$$D_{ii}^{-1} = \frac{1}{A_{ii}} \quad D_{ij}^{-1} = 0 \quad i \ne j \tag{9}$$

$$E_{ii} = 0 \quad E_{ij} = A_{ij} \quad i \ne j$$

then the iteration loop equation set is simply:

$$x^0 = D^{-1} \cdot b, \quad x^{i+1} = D^{-1} \cdot (b - E \cdot x^i) \tag{10}$$

and convergence to the solution x^n is reached when the difference between successive solutions is very small:

$$(x^n - x^{n-1}) \cdot (x^n - x^{n-1}) < e^2 \tag{11}$$

where e is the specified tolerance, set to 1×10^{-10} in the SEIDEL solver developed for HST3D. For an N node simulation, the SEIDEL algorithm requires storage for N extra elements (the x^{i-1} array). Since within HST3D the DCG and SEIDEL algorithms are called alternately to solve the flow and transport equations, the total storage required for matrix solution is 7N (A matrix) + 2N = 9N elements.

3 THE PRE- AND POST- PROCESSOR

To start a particular simulation run, HST3D reads an input data file that contains all information about the simulation (such as dimensions, grid spacing, boundary conditions, hydraulic zones, wells, and time steps). This file has a fixed format that is not user friendly. A simple menu driven preprocessor was developed that can be used to construct simulations involving constant head boundary conditions at opposite ends of the simulation region, no flow boundary conditions on the bottom and sides of the region, and either an unconfined or confined boundary condition on the top. The pre-processor allows interactive modification of the various model parameters and elementary parameter checking before generating the input data file. It also allows input of observed field data from monitoring wells, for comparision with simulated results. All model parameter information and field data is prepended to the input data file, so that the complete configuration of any simulation run can be recovered by the pre-processor directly from the file. The HST3D program has been modified to transfer the parameter information from the input file to the beginning of the output file, without modification, for use by the graphics post-processor. The preprocessor program is written in K&R C language (Kernighan & Ritchie, 1978) and is 4069 lines long.

HST3D simulation results are written to an ASCII format output file. Within each file, the spatially varying parameters (e.g. flow velocity, head, or solute concentration) are divided into a series of two dimensional node slices, and every slice is printed as a table of numbers. Each number in a table represents the value of the parameter at a particular node at a particular time. For simulation runs with large numbers of nodes (N ~ 100K) large output files (over 100 Mb) can be generated. Manual interpretation of the output file becomes impractical in these cases. A post-processor capable of generating graphical displays of the velocity, head, solute and well information in the output file is needed. Previously available general format graphics postprocessors were found to be too slow when handling large HST3D model output files. An interactive, X Window based, HST3D-dedicated post-processor was developed that also allows graphical comparison between different simulation runs and comparison of simulation results with observed field values. The display device is assumed to be an X Window version 11.4 server (Jones, 1989). The post-processor program is written in K&R C language and is 10,318 lines long. It is a derivative of the FX graphics program (Laslett, 1994).

4 THE SIMULATION REGION

The simulation region consists of an area surrounding a pumping well P1 (see Fig. 1). The X direction grid of the simulation is aligned from west to east, extending 100 m west and 150 m east of P1, with X = 0 m at the western boundary and X = 250 m

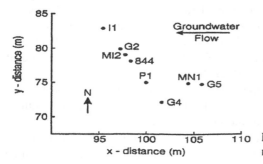

Figure 1. Field location of boreholes, with model scale.

at the eastern boundary. The Y direction grid is aligned from south to north 75 m either side of P1, with Y = 0 m at the southern boundary and Y = 150 m at the northern boundary. The bottom of the simulated region is set at a horizontal level 7 m below the Australian Height Datum (AHD), corresponding to the base of the unconfined aquifer. As this aquifer is unconfined, the top of the simulated region changes in space and time and does not have a constant vertical level. The vertical level of the uppermost layer of nodes is set at 0 m AHD. The Z direction grid is aligned vertically upwards with Z = –7 m at the bottom of the region and Z = 0 m at the top. Thus the origin of the model co-ordinate system (X = 0, Y = 0, Z = 0) is at the south-west corner and at the AHD datum level.

To model fine detail head and flow structure, the X and Y grid spacings are set from 0.1 m to 0.5 m apart within this area. The Z grid spacing is generally set at 0.25 m in the upper 1 m, and 0.5 m in the next 1 m. At greater depth (below Z = –2 m) the Z grid spacing is set to 1 m or 2 m.

The southern (Y = 0 m), northern (Y = 150 m) and bottom (Z = –7 m) boundaries of the simulation region are set to no-flow boundary conditions. The background water flow through the site is approximated by imposing constant head conditions on the western (head = –0.34 m AHD at X = 0 m) and eastern (head = 1.01 m AHD at X = 250 m) boundary surfaces of the simulation, resulting in an east to west water flow. The initial head gradient is 5.4 m/km.

A number of intensive hydraulic parameter values were used in all simulation runs. In particular, the values for total aquifer porosity (or saturated water content q_s) of 0.45, porous medium bulk density (r_b) of 1400 Kg/m³, and NAPL volumetric content of 0.073 close to the water table were chosen on the basis of experimental results (Johnston & Patterson, 1994). Note that in the HST3D code, the specific yield (when using the unconfined aquifer model) is assumed to equal the aquifer effective porosity, which in turn is assumed to equal the total porosity.

5   THE CONFINED AQUIFER APPROXIMATION

A 'free surface' boundary condition imposed at Z = 0 m allows a simulated unconfined aquifer to partially fill and drain. However, a major limitation of HST3D is that the water level must not drop below the top layer of cells, or rise more than one cell thickness above the top layer. Therefore if large changes in water level occur, the top layer must be very thick. To contain the full range of water levels occurring in the simulated region, a top-layer cell thickness of at least 1 m is required, enabling the water level in any top cell to fluctuate from Z = –1 to 1 m. Examination of fine scale structure near the water table is thus impossible, because the values of head, velocity and solute concentration are averaged over the entire cell. This is addressed below.

Generally, within top-layer cells of an unconfined aquifer simulation, the rate of change of water volume with head is equal to the specific yield times the surface area. The specific yield represents the capacity to drain and fill the pore space of the cell and is always less than or equal to the porosity. In all cells of a confined aquifer simulation (and any subsurface cell in an unconfined aquifer simulation), the rate of change of water volume with head is equal to the specific storage times the cell vertical thickness times the surface area, or the storage coefficient times the surface area

(Freeze & Cherry, 1979). The specific storage represents the compressibility of the fluid and porous medium in the cell.

Now consider a confined aquifer system split into two hydraulic zones, with the top zone covering the same volume as the thick top layer of cells of the unconfined aquifer system, and the bottom zone covering the rest of the system. If the storage coefficient of the top zone is set to the same value as the specific yield of the unconfined aquifer, and the specific storage of the bottom zone to the same value as the specific storage of the unconfined aquifer, then the two systems should exhibit similar qualitative behaviour. The top zone represents the high water storage capacity of the pore space and its capacity to drain and fill at the water table. The bottom zone represents the low compressibility of the saturated medium below the water table.

The advantage of the confined aquifer system is that the top zone can be made several cell layers thick, enabling examination of fine detail near the water table. A possible drawback is that the sensitivity of porosity to changes in head is dependent on the storage coefficient. However, it can be shown that an extremely low head does not cause a negative porosity in a cell, but changes can be significant. Also, since in the unconfined system the cells are partially filled with water, unlike the confined system, the Z coordinate grid is translated and deformed when mapped from the confined system to the unconfined system. This can, however, be accounted for.

Five simulation runs were used to assess the validity of using a confined aquifer model to simulate the unconfined sandy aquifer. The common parameters for each run are: number of grid lines in the X direction (NX) = 56, number of grid lines in the Y direction (NY) = 53, X direction grid spacing within the well field (Δx) = 0.5 m, Y direction grid spacing within the well field (Δy) = 0.5 m, hydraulic conductivity in the X, Y, and Z directions (K) = 7.41 m/day, and background water flow velocity (BV) = 0.124 m/day. BV is calculated from Darcy's law. Tables 1, 2 and 3

Table 1. Type and layering.

Simulation	NZ	Aquifer type	Upper 2 m
DZCF	12	DZC	Fine
DZCC	7	DZC	Coarse
SZC1	12	SZC	Fine
SZC2	12	SZC	Fine
SZUC	7	SZU	Coarse

NZ = no. of grid lines in the Z direction. DZC = Double Zone Confined, SZC = Single Zone Confined, and SZU = Single Zone Unconfined. Fine = 7 layers in the upper 2 m and Coarse = 1 layer in the upper 2 m of the aquifer.

Table 2. Porosities.

Simulation	n_L	n	n_U
DZCF	0.322		0.322
DZCC	0.322		0.322
SZC1		0.322	
SZC2		0.322	
SZUC		0.322	

n, n_U and n_L = effective porosity of the single zone, upper zone and lower zone respectively.

Table 3. Storage coefficients.

Simulation	SC_L	SC	SC_U
DZCF	1.703×10^{-4}		0.322
DZCC	1.703×10^{-4}		0.322
SZC1		2.834×10^{-4}	
SZC2		0.322	
SZUC		2.834×10^{-4}	

SC, SC_U and SC_L = confined storage coefficient of single zone, upper zone and lower zone respectively.

Table 4. Calibration assumptions for MI2.

Assumptions	Match at MI2
Experimental value for Porosity	$n_U = 0.45$, $n_L = 0.45$
Specific Yield = Porosity	$SC_U = 0.45$
	$K = 15$ m/day
RWMI2	$BV = 0.18$ m/day
Porosity can change	$n_U = 0.32$, $n_L = 0.32$
Specific Yield = Porosity	$SC_U = 0.32$
Waterlevel match at 844,G4	$K = 7.41$ m/day
RNMI2	$BV = 0.12$ m/day
Experimental value for Porosity	$n_U = 0.45$, $n_L = 0.45$
Specific Yield < Porosity	$SC_U = 0.2$
Waterlevel match at 844,G4	$K = 11.115$ m/day
RSMI2	$BV = 0.13$ m/day
Upper Zone NAPL reduced Porosity	$n_U = 0.38$, $n_L = 0.45$
Specific Yield < Porosity	$SC_U = 0.26$
Waterlevel match at 844,G4	$K = 10$ m/day
NNMI2	$BV = 0.12$ m/day*

* Background velocity calculated for lower zone

display individual parameter values for each run. The chosen values of K, n, n_L and n_U were taken from simulation run RNMI2 (see Section 6, and Table 4).

Results from all of the confined aquifer runs were compared with those of the unconfined aquifer run (SZUC) to see which behaved most similarly. SZC1 has a confined storage coefficient (SC) equal to the SC of SZUC. SZC2 has a SC equal to the porosity (n) of SZUC. Neither of these single zone confined systems matched the unconfined system well. For example, Z velocity components at position ($X = 95$, $Y = 75$, $T = 7.5$) were negative while for the unconfined aquifer system (SZUC) velocities were positive. Major differences in head were also observed between SZUC and the confined simulations. At position ($X = 96$, $Y = 77$, $T = 7.5$) head decreases with depth in the single zone confined aquifer simulations, while for SZUC, head increases with depth.

In contrast, the double zone confined system (DZCF) matched the shape of the head depth profile of SZUC and was accurate to within 0.013 m at this position. At

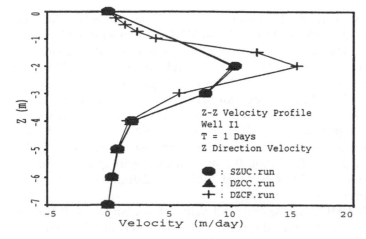

Figure 2. Velocity depth
profile at I1 (T = 1 day).

positions far from wells P1 and I1 the match between heads and velocities is good. The greatest mismatch occurs close to these wells. The largest head difference of 0.228 m occurs at well I1 at $T = 3.5$ days. The largest velocity magnitude difference of 4.84 m/day also occurs at I1 at $T = 1$ day (Fig. 2). Figure 2 shows the mismatches to be pronounced in the zone of grid refinement (upper 2 m) indicating that the head and velocity differences at the wells is largely due to different vertical grid spacing rather than intrinsically different behaviour between the two systems. DZCC has the same vertical grid spacing as SZUC, reducing the largest head difference to 0.087 m, and the largest velocity magnitude difference to 0.36 m/day, both at I1 at $T = 3.5$ days. The peak Z direction velocity of DZCC also occurs at the same depth as SZUC. On the basis of these results, it seemed justifiable to use a double zone confined aquifer system to simulate the unconfined aquifer at the field site.

6 MODELLING THE TRACER TESTS

The use of centred in space and centred in time finite differencing caused insurmountable instability in the solutions of the flow and transport equations. The use of upstream and backward in time differencing ensured stability but led to greatly increased numerical dispersion (generated by numerical roundoff error and finite grid spacing). Even with both longitudinal and transverse dispersivity coefficents set to zero in all simulation runs, numerical dispersion of the solute plumes far exceeded any observed hydrodynamic dispersion of the solute in the field. This meant that an exact match between simulated and observed solute curves was unachievable. Numerical dispersion can be decreased by refining the model grid, but due to current limitations in computing resources, any further refinement of the grid was not practical.

In the calibration runs, the sandy unconfined aquifer was simulated by using the double zone confined aquifer system. Tracer was applied by adding water into the zone $Z = -0.25$ to 0 m in wells G2 and G5 at a rate of 0.015 m³/hr for 0.5 days, starting at $T = 3.5$ days, with a tracer concentration of 300 mg/L. Breakthrough was observed in the field approximately 2 days after the end of injection at both MI2 and MN1. MI2 is 0.96 m south-south-east from G2, and MN1 is 1.4 m west of G5.

Eight different simulation runs were generated for which K and SC_U values were adjusted to match observed and simulated breakthrough times in wells MN1 and MI2, using different assumptions about the hydrological properties of the aquifer. Being able to match simulated head curves with field observations of water levels in wells I1, P1, G4, and 844 was considered to be an important criterion for the success of each run. The common parameters for all eight runs are: NX = 108, NY = 91, NZ = 10, double zone confined aquifer system, fine upper zone vertical grid spacing, 0.5 m horizontal grid spacing over field plot, 0.1 m horizontal grid spacing near G2, MI2, G5 and MN1, and $SC_L = 1.703 \times 10^{-4}$ (see Table 3 for definition). Tables 4 and 5 display individual parameters of each run.

For RWMN1 and RWMI2, the experimental value of porosity ($n = 0.45$) was used, it was assumed that the unconfined specific yield equalled n, and the value of K was manipulated. The above assumptions mean that SC_U must equal n. It was found that no single value of K could match the breakthrough times at both wells, suggesting a non-uniform aquifer. The breakthrough time was matched at MI2 for $K = 15$ m/day (RWMI2) and at MN1 for $K = 30$ m/day (RWMN1). The simulated water level fluctuations at 844 and G4 were significantly less than the observed fluctuations.

For RNMN1 and RNMI2, the experimental value of porosity ($n = 0.45$) was abandoned in favour of achieving a better water level match at wells 844 and G4, whilst still maintaining the assumption that specific yield equals n. The water levels at 844 and G4, and the bromide breakthrough time at MI2, were matched for $n = 0.322$ and $K = 7.41$ m/day (RNMI2). The water levels at 844 and G4, and the bromide breakthrough time at MN1, were matched for $n = 0.23$ and $K = 10.29$ m/day

Table 5. Calibration assumptions for MN1.

Assumptions	Match at MN1
Experimental value for Porosity	$n_U = 0.45$, $n_L = 0.45$
Specific Yield = Porosity	$SC_U = 0.45$
	$K = 30$ m/day
RWMN1	$BV = 0.36$ m/day
Porosity can change	$n_U = 0.23$, $n_L = 0.23$
Specific Yield = Porosity	$SC_U = 0.23$
Waterlevel match at 844,G4	$K = 10.29$ m/day
RNMN1	$BV = 0.24$ m/day
Experimental value for Porosity	$n_U = 0.45$, $n_L = 0.45$
Specific Yield < Porosity	$SC_U = 0.07$
Waterlevel match at 844,G4	$K = 21$ m/day
RSMN1	$BV = 0.25$ m/day
Upper Zone NAPL reduced Porosity	$n_U = 0.38$, $n_L = 0.45$
Specific Yield < Porosity	$SC_U = 0.05$
Waterlevel match at 844,G4	$K = 20$ m/day
NNMN1	$BV = 0.24$ m/day*

* Background velocity calculated for lower zone

(RNMN1). In neither case did the water levels at I1 and P1 match. The observed water levels at I1 vary over a greater magnitude than the simulated water levels and the observed water levels at P1 are always lower than the simulated water levels, indicating possible well loss effects that have not been modelled.

For RSMI2 and RSMN1, the porosity ($n = 0.45$) is maintained but specific yield (and hence SC_U) is allowed to fall below n. The water levels at 844 and G4, and the bromide breakthrough time at MI2, was matched for $n = 0.45$, $SC_U = 0.2$ and $K = 11.11$ m/day (RSMI2). The water levels at 844 and G4, and the bromide breakthrough time at MN1, were matched for $n = 0.45$, $SC_U = 0.07$ and $K = 21$ m/day (RSMN1). Again, no satisfactory match was obtained at I1 and P1.

For NNMN1 and NNMI2, the presence of NAPL in the soil pore space near the water table within the field plot was accounted for by lowering n_U. The measured average volumetric NAPL content over the contaminated interval of 0.073 was used to give an estimate of the net volumetric water content (q) of 0.377. This value of q was used for n_U assuming that the volume occupied by the NAPL does not contribute to water flow. The water levels at 844 and G4, and the bromide breakthrough time at MI2, were matched for $SC_U = 0.26$ and $K = 10$ m/day (NNMI2). The water levels at 844 and G4, and the bromide breakthrough time at MN1, was matched for $SC_U = 0.05$ and $K = 20$ m/day (NNMN1). Figures 3 and 4 show the observed versus simu-

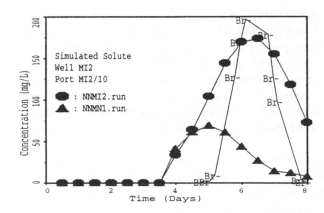

Figure 3. Tracer breakthrough curves at well MI2.

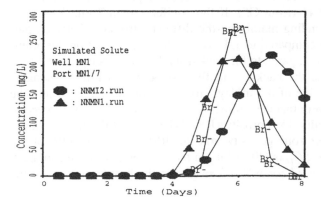

Figure 4. Tracer breakthrough curves at well MN1.

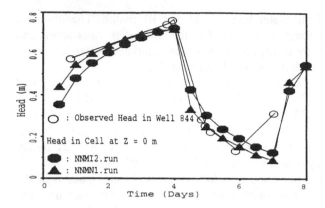

Figure 5. Head time history at well 844.

lated bromide tracer curves, and Figure 5 shows a typical observed versus simulated water level, at well 844 in this case. Again, no satisfactory match was obtained at I1 and P1.

7 SUMMARY AND CONCLUSIONS

This paper deals mainly in the practicalities of using HST3D to model large simulations. New matrix solvers developed for HST3D succeeded in reducing the run time for the simulations presented in this paper from days to hours. Development of menu driven pre- and post-processors, and modification of the HST3D code to allow transfer of observation data to the output file, made direct on-screen comparison between simulated and observed results possible. The visualization tools provided by the post-processor proved to be an invaluable aid when dealing with simulations of this size and complexity. The figures in this paper were taken directly from the post-processor display window.

The double zone confined aquifer model can be used with reasonable accuracy to mimic the behaviour of the unconfined aquifer at the field site. However, the aquifer was apparently too heterogeneous to accurately calibrate a numerical model using a single set of hydrological parameters for the whole region. One indication of this is that the arrival time of solute at borehole MI2 (from G2) is approximately 2 days, the same as the arrival time at MN1 (from G5). If the aquifer was uniform the arrival times would be different, depending mainly on the different distances between piezometer and borehole (0.96 m compared to 1.4 m), and to some extent on the orientation of the solute travel path compared to the background flow velocity.

In all eight simulation runs, the background flow velocity is within reasonable proximity to the actual field value of 0.2 m/day. However, for those runs where the bromide breakthrough times were matched at MN1 and specific yield was allowed to fall below the experimental value for porosity (RSMN1 and NNMN1), SC_U values are less than 16% of the upper zone porosity, n_U. For RSMI2, SC_U is 44% of n_U, which is still low. For NNMI2, SC_U is 68% of n_U, a more acceptable value, suggesting that the parameter values used for NNMI2 might be the best to use in further simulations.

ACKNOWLEDGEMENTS

This work has been funded by a cooperative research agreement with Broken Hill Pty. Co. Ltd. Comments and assistance of Colin Johnston and David Herne are appreciated.

REFERENCES

Buckingham, R.A. 1962. *Numerical Methods*: 430-432. Sir Isaac Pitman & Sons.

Davis, G.B., Johnston, C.D., Thierrin, J., Power, T.R. & Patterson, B.M. 1993. Characterising the distribution of dissolved and residual NAPL petroleum hydrocarbons in unconfined aquifers to effect remediation. *AGSO Journal of Australian Geology and Geophysics* 14(2/3): 243-248.

Davis, G.B. & Salama, R.B. 1995. Modelling point sources of groundwater contamination: a review and case study. In: *Advances in Environmental Science, Groundwater Contamination*: 111-140. Science Reviews, Laws and Stimson, Surrey.

Freeze, R.A. & Cherry, J.A. 1979. *Groundwater*: 58-62. Prentice Hall.

Johnston, C.D. & Patterson, B.M. 1994. Distribution of nonaqueous phase liquid in a layered sandy aquifer. *Hydrocarbon Bioremediation*: 431-437. Lewis.

Jones, O. 1989. *Introduction to the X Window System*. Prentice Hall.

Kernighan, B.W. & Ritchie, D.M. 1978. *The C Programming Language*. Prentice Hall.

Kipp, K.L. 1987. *HST3D: a computer code for simulation of heat and solute transport in three dimensional ground-water flow systems*. US Geological Survey, Denver, Colorado.

Laslett, D. & Davis, G.B. 1993. Modelling three-dimensional flow and solute transport for the in situ bioremediation evaluation trial, CSIRO Division of Water Resources Report to BHP, No. 93/34.

Laslett, D. 1994. FX: A three-dimensional graphical visualisation tool and macro language users manual. CSIRO Division of Water Resources Divisional Report, No. 94/1.

Meyer, P.D., Valocchi, A.J., Ashby, S.F. & Saylor, P.E. 1989. A numerical investigation of the conjugate gradient method as applied to three-dimensional groundwater flow problems in randomly heterogeneous porous media. *Water Resources Research* 26(6): 1440-1446.

Press, W.H., Flannery, B.P., Teukolsky, S.A. & Vettering, W.T. 1987. *Numerical recipes: The art of scientific computing* 54: 70-71, 295-297, 301-307. Cambridge University Press.

CHAPTER 18

Density induced flow and solute transport below a saline lake bed

KUMAR A. NARAYAN & CRAIG T. SIMMONS
Centre for Groundwater Studies and CSIRO Division of Water Resources, Glen Osmond, Australia.

ABSTRACT: The movement of salt from a saline lake bed/evaporation basin to the underlying groundwater system is investigated. The density dependent flow behaviour is modelled in cross section using a 2-D finite element model SUTRA. Due to the inherent salinity contrast between the saline lake brine and groundwater, the system is characterised by a state of density stratification which is the major cause of the resulting instability. Mixed convection flow (hydraulically and buoyancy-driven) occurs which subsequently controls the solute concentrations below the lake bed in order to achieve a stable density gradient (Rayleigh convection). This type of flow regime is particularly relevant to saline disposal basins in the Murray-Darling Basin of Australia. Numerical examples simulating 2-D brine convection from the lake bed salt are given. The conditions which define the onset of such convective phenomena are investigated. A stability criteria based on the theory of Rayleigh convection is employed and is seen to be an extremely powerful tool for predicting the long-term behaviour of saline disposal basins.

1 INTRODUCTION

The Murray-Darling Basin of Australia is characterised by its many salt lakes and evaporation basins. These have been studied by numerous researchers including Macumber (1991). There are currently more than 180 disposal basins throughout the Murray-Darling Basin which receive in the order of 100 million cubic metres of disposal water annually. Each year they divert around one million tonnes of salt which would otherwise directly reach the Murray River. However, concerns have been expressed about the possibility of return losses from these evaporation basins to the River Murray by transport in the groundwater system.

Allison & Barnes (1985) have pointed out the significance of the accumulated salt budgets observed for various salt lakes. It is clear that the amount of salt present in these lakes is considerably less than would be expected for terminating points of drainage systems. This salt deficit is also observed in the salt budgets of evaporation basins. This may be attributed to the slow downward convection of dense saline water

221

(brine) beneath these salt lake/evaporation basin beds which could partially account for the low salt budgets as suggested by Teller et al. (1982). In situations where a saline lake bed overlays relatively fresh (and consequently less dense) groundwater, there is a tendency for the dense saline solution to move downward thereby displacing the less dense fresh water upwards. This results in free convective currents which can affect the underlying groundwater regime.

Density induced fluid flow may also be caused by variations in temperature and pressure. The phenomenon of convective currents arising from temperature variations within a flow domain are well studied and understood. Wooding (1963) has studied the effect of convection in saturated porous media under nonisothermal conditions. However, mixed convection flow (hydraulically and buoyancy-driven transport) which occurs due to inherent salinity contrasts in the saline lake bed/groundwater system and external head gradients are not well understood.

The theoretical formulation of variable density flow has been addressed by a number of authors including Bear (1972). A US Geological Survey model SUTRA (*Saturated-Unsaturated TRA*nsport) developed by Voss (1984) takes into account variable density flow and solute transport in subsurface systems. The SUTRA model has been previously used for numerical simulation of density dependent flow including work by Voss & Souza (1987), Souza & Voss (1987), Ghassemi et al. (1990) and Narayan & Armstrong (1995). Herbert et al. (1988) used the NAMMU model to simulate solute transport in a subsurface environment containing a salt dome formation. We firstly modelled a salt dome structure using SUTRA to synthesise the results of Herbert et al. (1988) before proceeding to evaporation basin modelling.

SUTRA has been used in this investigation to show that systems in which dense saline water overlies less dense groundwater are characterised by the development of an instability, which superimposes perturbations in concentration distribution on the mixed flow regime. This motion is then interpreted as being caused by two driving forces: One arising from piezometric head differences and the other arising from a buoyancy force acting vertically upward. Thus, for such systems instability development causes the fluids to mix in order to achieve a stable density gradient (Rayleigh convection). A stability criterion developed by Simmons & Narayan (1997) for mixed convective flow systems is used in this study. The relationships between Rayleigh number, instability and salt front movement from the disposal basin have been established by way of numerical example.

2   GOVERNING THEORY AND EQUATIONS

The governing equations for fluid mass and salt conservation provide a starting point for the theoretical background. The Rayleigh number is also briefly introduced.

2.1   *Mass conservation*

The fluid mass balance is:

$$\frac{\partial(\varepsilon\rho)}{\partial t} = -\nabla \cdot (\varepsilon\rho V) + Q_p \tag{1}$$

where $\varepsilon(x, y, t)$ is a dimensionless porosity, $\rho(x, y, t)$ is the fluid density, $V(x, y, t)$ is average fluid velocity, $Q_p(x, y, t)$ is a fluid mass source, x and y are coordinate variables, t is time and ∇ is the usual gradient operator.

Density is given as a linear function of concentration:

$$\rho = \rho_0 + \frac{\partial \rho}{\partial C}(C - C_0) \tag{2}$$

where ρ_0 is the fluid density at a base concentration C_0 and $\partial \rho / \partial C$ is a constant coefficient of density variability.

In quantitative terms the implied coupling between flow and salinity in some confined enclosure requires a form of Darcy's law that includes both pressure and density forces. With variable fluid density, the fluid flow equation is expressed in terms of the pressure variable since the potential head function does not exist. The pressure gradient form of Darcy's law gives the mass average fluid velocity at any point in a cross section as:

$$V = -\left(\frac{k}{\varepsilon \mu}\right) \cdot (\nabla p - \rho g) \tag{3}$$

where $p(x, y, t)$ is the fluid pressure, g is the gravity vector, μ is fluid dynamic viscosity and $k(x, y)$ is the intrinsic permeability tensor.

2.2 Salt conservation

For a single species stored in solution, the solute mass balance equation may be expressed as:

$$\frac{\partial(\varepsilon \rho C)}{\partial t} = -\nabla \cdot (\varepsilon \rho V C) + \nabla \cdot [\varepsilon \rho (D_0 I + D) \cdot \nabla C] + Q_p C^* \tag{4}$$

where D_0 is the apparent molecular diffusivity in a porous medium of solutes in solution, I is the dimensionless identity tensor, C^* is the concentration of fluid sources expressed as a mass fraction. Bear (1972) has formulated the components of the mechanical dispersion tensor D to account for both transverse and longitudinal dispersivities respectively.

2.3 Rayleigh number analysis

Consider the balancing forces or fluxes in the subsurface system. Piezometric head differences give rise to the usual Darcian or advection flow. The density difference between the groundwater (concentration = C_0) and the saline lake fluid of higher concentration C_L gives rise to a vertical buoyancy force. The Rayleigh number indicates the likelihood of free convection occurring due to these buoyancy forces and is defined as the ratio of the buoyancy forces tending to cause flow to other forces tending to resist flow. Simmons & Narayan (1997) showed that the stability of mixed convective flow below a saline disposal basin was related to the magnitude of at least two nondimensional numbers, a Rayleigh number and a modified Peclet number.

Both of these parameters are defined in terms of basin scale hydrogeologic parameters. The stability criteria (Simmons & Narayan, 1997) is stated as:

$$Ra^* = \frac{Ra}{1 - Pe^*} \leq 1250 \qquad (5)$$

Here, the Rayleigh number Ra is defined by

$$Ra = \frac{U_c H}{D_T} = \frac{gk\beta (C_{max} - C_{min}) H}{\varepsilon v_0 (D_0 + \alpha T V_{amb})} = \frac{\text{Bouyancy forces}}{\text{Resistance forces}} \qquad (6)$$

where U_c is the convective velocity, H is the depth of the porous layer, D_T is the transverse dispersion coefficient, D_0 is the molecular diffusivity, g is acceleration due to gravity, k is the intrinsic permeability, $\beta = \rho_0^{-1} (\partial \rho / \partial C)$ is the linear expansion coefficient, C_{max} and C_{min} are the maximum and minimum values of concentration respectively (expressed as solute weight relative to weight of solution), ε is the aquifer porosity, $v_0 = \mu_0/\rho_0$ is the kinematic viscosity of the fluid, α_T is the transverse dispersivity and V_{amb} is the ambient velocity due to external head gradients.

The modified Peclet number Pe^* (Simmons & Narayan, 1997) is used to account for dispersion due to ambient groundwater flow and is defined as:

$$Pe^* = \frac{V_{amb} \alpha_L}{D_0 + V_{amb} \alpha_L} \qquad (7)$$

In the case where the ambient velocity V_{amb} is assumed negligible, free convection dominates and Pe^* is negligibly small. The modified Rayleigh number Ra^* becomes the simplified version used by Wooding (1989). In many field based examples, ambient groundwater flow and mechanical dispersion cannot be ignored and the criteria above must be used to account for velocity dependent dispersion.

Previous work (Wooding et al., 1997a,b) has shown that instabilities increase the efficiency of salt or heat transfer from either a disposal basin or heat source. The dimensionless number used to describe salt fluxes from the basin to the underlying aquifer system is the Nusselt number. For a two-dimensional porous medium bounded by length L_B and width unity, the Nusselt number for the saline disposal basin can be represented as:

$$Nu = \frac{Q}{D_0 \Delta \rho L_B} \qquad (8)$$

where Q is the solute flux rate, D_0 is the molecular diffusivity, $\Delta \rho$ is the maximum density gradient across the saline disposal basin, and L_B is the length of the saline disposal basin. The Nusselt number is a useful stability indicator because it directly reflects the amount of salt leaving the disposal basin.

3 NUMERICAL MODELLING

In order to obtain solutions to our problem, both mass and salt conservation partial differential equations must be solved numerically. This requires appropriate spatial and temporal discretisation together with relevant initial and boundary conditions.

3.1 *The evaporation basin model*

Mathematically, an idealised (hypothetical) representation of an evaporation basin is illustrated in Figure 1. The field is defined as a vertical, two-dimensional flow in a homogenous and isotropic porous medium of depth H and horizontal extent L. The geometry of the evaporation basin has been greatly simplified to a hypothetical model so that attention is not diverted from the concept being studied – the physical phenomena of Rayleigh instability. The region in Figure 1 has horizontal extent $L = 9000$ m and vertical extent $H = 90$ m. The evaporation basin is located on the surface of the model at $2900 < x < 3900$ m. This spatial domain was discretised to form a completely uniform mesh containing 4646 nodes and 4500 elements. This mesh was generated using FEMCAD software developed by Beer & Mertz (1990) and then converted to the required SUTRA format using an interface package developed by Buia et al. (1994). This software allowed for rapid, user friendly mesh generation. The elemental size was such that $\Delta x = 90$ m and $\Delta y = 2$ m. This was satisfactory for our requirements because the flow regime studied in this paper is dominated by a buoyancy generated flow thus requiring finer discretisation in the vertical direction.

3.2 *Boundary conditions*

The aquifer underlying the evaporation basin is assumed to be completely homogeneous and isotropic. In order to simulate the effect of recharge and discharge zones, a water table that sloped linearly downwards was modelled in cross section. This was achieved by employing a top boundary condition which has pressure varying linearly on the top surface from $10^5\,Pa$ on the left to $0\,Pa$ on the right, where $p = 10^5\,(9000-x)/9000\,Pa$. The remaining boundaries of the model are taken to be impermeable to water flow and are therefore specified as no flow boundaries. The boundary conditions for salt transport are specified by an evaporation basin of concentration C_L at

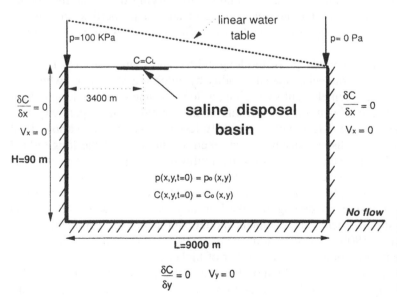

Figure 1. Simulation geometry, boundary and initial conditions for an evaporation basin system.

2900 m < x < 3900 m on the top surface. This concentration is specified for varying model simulations. The base and the vertical side walls are taken to be impermeable to solute.

3.3 Initial conditions

The choice of pressures and concentrations used for initial conditions are critical for accurate and meaningful results to be obtained. To represent initial conditions for the system, a steady-state run was made with pressures along both ends being defined by the hydrostatic equation. Namely, pressure variations given by $p = \rho_{gw}gh$ were imposed along the vertical boundaries where p is the hydrostatic pressure, ρ_{gw} is the density of groundwater (which initially occupies the aquifer system), g is acceleration due to gravity and h is the depth below the system surface. Thus, the pressure at the top of the left hand boundary is 10^5 Pa and increases linearly with depth. Similarly the pressure at the top of the right hand side boundary is zero and increases linearly with depth. A steady state simulation was run with these initial conditions for pressure and groundwater concentrations of 5000 mgL^{-1} TDS throughout the model for initial concentrations. The resulting steady state pressures were used as initial conditions for all transient simulations along with an initial groundwater concentration of 5000 mgL^{-1} TDS throughout the model.

3.4 Model parameters

It has been seen that the Rayleigh number of convective flow is dependent upon a number of parameters. These include soil parameters such as intrinsic permeability and porosity. The magnitude of the Rayleigh number is also dependent upon kinematic viscosity, molecular diffusivity, the depth of the porous medium and the relative concentration difference driving the buoyancy force. Thus it is obvious that varying system Rayleigh numbers can be obtained by varying these parameters accordingly. We performed simulations for a range of aquifer properties (permeabilities and porosities) and a range of concentration differences between the basin and groundwater.

SUTRA also requires knowledge of the longitudinal and transverse dispersivities. Voss (1984) suggests guaranteeing spatial stability by enforcing $\nabla_L \leq 4\ \alpha_L$, where ∇_L is the local distance between element sides along a flow line and α_L is the longitudinal dispersivity. Given the dimensions of our mesh, the longitudinal dispersivity was taken as 50 m. Transverse dispersivity is typically less well known for various field problems and transverse dispersivity has been taken as one-tenth of the longitudinal dispersivity, namely 5 m. Thus, it is obvious that in this model type, the dispersivity is more related to the mesh size than to the system under analysis, but a concerted attempt was made to use physically reasonable values for dispersion. A sensitivity analysis of solute transport modelling at Lake Ranfurly by Charlesworth & Narayan (1994) was of assistance here. The complete model parameters for a moderate Rayleigh number of $Ra = 1800$ and Peclet number $Pe^* = 0.98$ are given in Table 1.

Given also are results for a Rayleigh number of higher order $Ra = 5500$ and Peclet number $Pe^* \approx 1$. This was achieved by modification of the aquifer properties such that intrinsic permeability $k = 1.05 \times 10^{-12}$ m^2 and porosity $\varepsilon = 0.3$. The concentration dif-

Table 1. Model Parameters for $Ra = 1800$ and $Pe^* = 0.98$.

Freshwater density $(\rho) = 1000$ kgm^{-3}
Groundwater concentration = 5000 mgL^{-1}
Lake concentration $(C_L) = 200,000$ mgL^{-1}
Fluid dynamic viscosity $(\mu) = 10^{-3}$ kgm^{-1}s^{-1}
Coefficient of fluid density change $(\partial\rho/\partial C) = 700$ kgm^{-3}
Water compressibility = 4.5×10^{-10} Pa^{-1}
Aquifer (solid matrix) compressibility = 1×10^{-8} Pa^{-1}
Porosity $(\varepsilon) = 0.35$
Intrinsic permeability $(k) = 9.76 \times 10^{-14}$ m^2
Longitudinal dispersivity $(\alpha_L) = 50$ m
Transverse dispersivity $(\alpha_T) = 5$ m
Molecular diffusivity $(D_0) = 2.8 \times 10^{-9}$ m^{2}s^{-1}
Acceleration due to gravity $(g) = 9.81$ ms^{-2}

ference between the evaporation basin and the subsurface groundwater was set to be 0.5 kg kg^{-1} TDS, therefore giving a much higher lake concentration of 505,000 mgL^{-1}. It should be noted that this concentration value exceeds solubility limits but is used as an artifact to produce instabilities that demonstrate the capability of the numerical model. All other model parameters are as given in Table 1 and were not modified. Note also that the hydraulic conductivity K may be obtained simply using $K = k\rho g/\mu$, with all parameters as defined in Table 1.

4 RESULTS AND DISCUSSIONS

Presented here are the numerical results for $Ra = 1800$ and $Pe^* = 0.98$. Also given are results for $Ra = 5500$ and $Pe^* \approx 1$. These offer insight into the groundwater flow and solute transport below an evaporation basin and in particular highlight the downward fingering of salt solution from the evaporation basin. By variation of soil and basin parameters in turn we were able to modify the numerical value of the convective Rayleigh number and therefore 'observe' the relative impact on groundwater flow and solute transport that each Rayleigh number had.

The first simulation shown in Figures 2a, b, c and d depict the slow downward convective mixing of dense saline water beneath the salt lake beds at 12 years, 25 years, 52 years and 94 years respectively in cross sectional view. In the time span shown there is very little lateral movement of salt in the aquifer and therefore the movement of the salt front towards the discharge zone on the right of the evaporation basin model is negligible. It can be seen that once lobe-shaped instabilities are sufficiently established, they penetrate into the aquifer. Buoyant plumes of less dense groundwater must rise to replace the liquid in these less dense fingers. It is clear that this system is unstable and this is also evident by evaluation of Equation (5), which gave $Ra^* \sim 10^5$. Velocity vectors for this simulation are given in Figure 3. This clearly shows the free convection flow associated with the Rayleigh instability phenomena.

The second simulation (Fig. 4) has an increased basin salinity level which gives rise to a larger concentration gradient. The buoyancy force is therefore greater in this case than the previous example. The corresponding modifications to aquifer permeability and porosity allow for a much higher Rayleigh number to be achieved. It can be seen that the higher Rayleigh number (rapid mixing) of $Ra = 5500$ gives rise to much faster convective motion in the aquifer (~100 times faster than the previous Rayleigh simulation). This increase in convective velocity U_c means downward transport of saline solution is also correspondingly faster. This is easily seen in the cross sectional concentration profiles given in Figures 4a, b, c and d at 5, 25, 52 and 94 year operating times respectively.

A careful comparison between the concentration profiles of Figures 2 and 4 reveals a marked difference in the salt front movement. Once again, the onset of instability was predicted using the stability criteria (Eq. 5) which gave $Ra^* \sim 10^5$.

A third case was run with a much lower Rayleigh number of $Ra = 450$ and $Pe^* = 0.6$. This was obtained by using a much smaller (and physically unrealistic for an

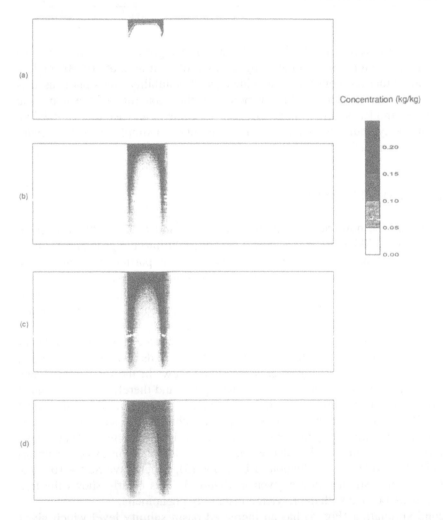

Figure 2. $Ra = 1800$, $Pe^* = 0.98$. Cross sectional distribution of computed concentration at: a) 5 years, b) 25 years, c) 52 years, and d) 94 years using 4646 nodes.

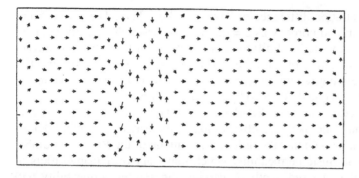

Figure 3. Cross sectional profile at 94 years for $Ra = 1800$ and $Pe^* = 0.98$. Convective velocity $\sim 10^{-8}$ ms^{-1} and advective velocity $\sim 10^{-9}$ ms^{-1}.

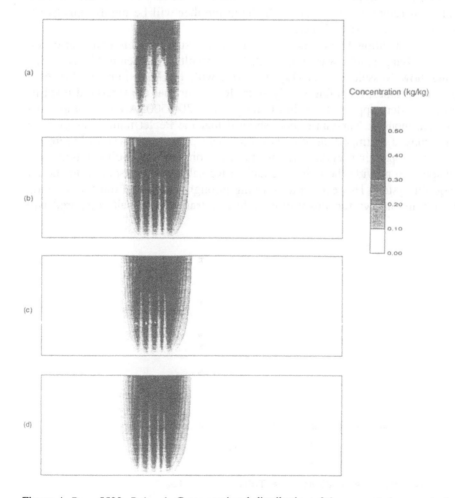

Figure 4. $Ra = 5500$, $Pe^* \approx 1$. Cross sectional distribution of the computed concentration at: a) 5 years, b) 25 years, c) 52 years, and d) 94 years using 4646 nodes.

evaporation basin) concentration difference of 0.0002 kg kg^{-1} TDS. The permeability was set to $k = 2 \times 10^{-12}$ m^2 and porosity as $\varepsilon = 0.2$. For this case, the density gradient was so small that the piezometric head differences resulted in a flow that was advection dominated. In fact evaluation of the criteria (Eq. 5) gave the value $Ra^* = 1100$ which is in the stable operating region. Free convective flow and instability were not observed in this case which is one dominated by *forced convection*.

Concentration differences in the range 0.1-0.3 kg kg^{-1}. TDS are typical for Murray-Darling disposal basins and salt lakes. By observation of the Rayleigh number definition in Equation (6), it can be seen that the order of magnitude of the Rayleigh number is highly sensitive to the intrinsic permeability which may vary over many orders of magnitude. For any aquifer of fixed depth H, varying the aquifer permeability will have a significant effect on the resultant convective Rayleigh number. With our concentration range relatively constricted to the above range, the buoyancy force is not highly variable and concentration differences alone will not vary the Rayleigh number dramatically. Therefore, the convective Rayleigh number will be highly sensitive to changes in the aquifer resistive forces.

By performing additional simulations, the effect of system Rayleigh number on salt flux to the underlying aquifer was monitored. These results are given in Figure 5. This figure shows how Nusselt number (Eq. 8) varies with Rayleigh number (Eq. 6) at various operating times, including steady-state. It is clear that the threshold level for stability in the system appears to be in the range $Ra \sim 300\text{-}500$. A more complicated analysis by Simmons & Narayan (1997) showed that the Peclet number as given in Equation (7) played an important role also. The main point to observe in Figure 5 is that there are two operating regimes. In the stable region, molecular diffusion and mechanical dispersion provide the only mechanism for salt to be transported to the underlying aquifer system. However, after moving through the critical transition region, convective mixing allows for vastly more effective transport of salt compared with

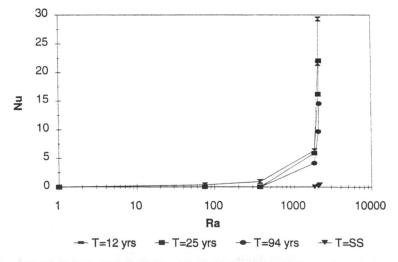

Figure 5. Nusselt number *Nu* versus Rayleigh number *Ra* at various times of basin operation, $T =$ 12 years, $T = 25$ years, $T = 94$ years and $T =$ steady state.

dispersion/diffusion alone. Hence, the unstable operating regimes associated with much higher Rayleigh numbers are also associated with significantly higher Nusselt numbers.

5 CONCLUSIONS

Numerical simulation and experiments show that salt disposal in evaporation basins will have implications for groundwater circulation and solute transport and that the residual brine fluid produces density gradients which set in motion a free convection cell where the Rayleigh number of the system is sufficiently high. The characteristic soil properties and basin salinity level are shown to have important implications in determining the convective Rayleigh number of the system. Compared with stable behaviour, unstable plumes which develop below a saline disposal basin are able to transport much greater amounts of salt to underlying groundwater systems and cause a larger degree of mixing. If a basin remains stable, the only mechanism for salt transport to the underlying aquifer is molecular diffusion and mechanical dispersion. Our numerical results show that the stability of dense brine plumes is determined by factors which include, amongst others, the relative density difference between the basin brine and ambient groundwater, the horizontal ambient velocity, the aquifer permeability and longitudinal and transverse dispersivities.

It was demonstrated that a mixed convective flow regime experienced a transition from stable to unstable behaviour when Rayleigh numbers became sufficiently high. This was described by a nondimensional number which combined two dimensionless numbers, a Rayleigh number and Peclet number. Stability was obtained under conditions in which $Ra^* = Ra/(1 - Pe^*) \leq 1250$. This uniquely defined the onset of plume instability for homogeneous systems with both isotropic and anisotropic permeability and dispersion.

By clearly understanding the groundwater and salinity dynamics of saline disposal basins, it will be possible to properly manage them so that large quantities of salt are not released into the environment. The development of a stability criteria will provide one set of tools necessary for brine management schemes in the Murray-Darling basin of Australia as well as many other disposal complexes where density effects may be significant.

ACKNOWLEDGMENTS

The authors are thankful to Peter J. Dillon and Glen R. Walker for their review of this work. Author C.T. Simmons wishes to gratefully acknowledge Flinders University of South Australia and the Centre for Groundwater Studies for Postgraduate Scholarship funding.

REFERENCES

Allison, G.B. & Barnes, C.J. 1985. Estimation of evaporation from the normally 'dry' Lake Frome in South Australia, *J. Hydrol.* 78: 229-242.

Bear, J. 1972. *Hydraulics of groundwater.* New York: McGraw Hill.

Beer, G. & Mertz, W. 1990. A user-friendly interface for computer aided analysis and design in mining. *Int. J. Rock. Mech. Sci. & Geomech. Abstr.* 27(6): 541-552.

Buia, M., Simmons, C.T. & Narayan, K.A. 1994. Interfacing of FEMCAD generated finite element mesh with groundwater flow and solute transport model SUTRA. CSIRO Division of Water Resources Divisional Report, 94/4.

Charlesworth, A.T. & Narayan, K.A. 1994. Sensitivity analysis of solute transport modelling at Lake Ranfurly, Victoria. *CSIRO Tech. Mem.,* 94/4.

Ghassemi, F., Jakeman, A.J. & Jacobson, G. 1990. Mathematical modelling of sea water intrusion, Nauru Island. *Hydrological Proc.* 4: 269-281.

Herbert, A.W., Jackson, C.P. & Lever, D.A. 1988. Coupled groundwater flow and solute transport with fluid density strongly dependent upon concentration. *Water Resources Research* 24(1): 1781-1795.

Macumber, P.G. 1991. *Interaction between groundwater and surface systems in northern Victoria.* Department of Conservation and Environment, Victoria.

Narayan, K.A. & Armstrong, D. 1995. Simulation of groundwater interception at Lake Ranfurly Victoria incorporating variable density flow and solute transport. *J. Hydrol.* 165: 161-184.

Simmons, C.T. & Narayan, K.A. 1997. Mixed convection processes below a saline disposal basin. *J. Hydrol.* 194: 263-285.

Souza, W.R. & Voss, C.I. 1987. Analysis of an anisotropic coastal aquifer system using variable-density flow and solute transport simulation. *J. Hydrol.* 92: 17-41.

Teller, J.T., Bowler, J.M. & Macumber, P.G. 1982. Modern Sedimentation and hydrology in Lake Tyrell, Victoria. *J. Geol. Soc. Aust.* 29: 159-175.

Voss, C.I. 1984. SUTRA: A finite-element simulation model for saturated-unsaturated fluid density-dependent groundwater flow with energy transport or chemically reactive single-species solute transport. *US Geol. Surv. Water Resour. Invest. Rep.* 84-4369.

Voss, C.I. & Souza, W.R. 1987. Variable density flow and solute transport simulation of regional aquifers containing a narrow freshwater-saltwater transition zone. *Water Resources Research* 23(1): 1851-1866.

Wooding, R.A. 1963. Convection in a saturated porous medium at large Rayleigh number or Peclet number. *J. Fluid Mech.* 15: 527-544.

Wooding, R.A., 1989. *Convective regime of saline groundwater below a 'dry' lakebed.* CSIRO Centre for Environmental Mechanics Technical Report No. 27.

Wooding, R.A., Tyler, S.W. & White, I. 1997a. Convection in groundwater below an evaporating salt lake: 1. Onset of instability. *Water Resources Research* 33(6): 1199-1217.

Wooding, R.A., Tyler, S.W., White, I. & Anderson, P.A. 1997b. Convection in groundwater below an evaporating salt lake: 2. Evolution of fingers and plumes. *Water Resources Research* 33(6): 1219-1228.

Printed and bound by CPI Group (UK) Ltd, Croydon, CR0 4YY

23/10/2024

01777685-0009